T0155565

SpringerBriefs in Health Care Management and Economics

More information about this series at http://www.springer.com/series/10293

Christo El Morr • Hossam Ali-Hassan

Analytics in Healthcare

A Practical Introduction

Christo El Morr
School of Health Policy and Management
York University
Toronto, ON, Canada

Hossam Ali-Hassan
Department of International Studies
Glendon College, York University
Toronto, ON, Canada

ISSN 2193-1704 ISSN 2193-1712 (electronic)
SpringerBriefs in Health Care Management and Economics
ISBN 978-3-030-04505-0 ISBN 978-3-030-04506-7 (eBook)
https://doi.org/10.1007/978-3-030-04506-7

Library of Congress Control Number: 2018967216

This Springer imprint is published by the registered company Springer Nature Switzerland AG
The registered company address is: Gewerbestrasse 11, 6330 Cham, Switzerland

For Valentina and Alexi and For Hala, Liane, and Yasma

Preface

This book offers a practical guide to analytics in healthcare. The book does not go into details of the mathematics behind analytics; instead it explains the main types of analytics and the basic statistical tools used for analytics and gives an illustration of how algorithms work by providing one example for each type of analytics. This allows the readers, such as students, health managers, data analysts, nurses, and doctors, to understand the analytics background, their types, and the kind of problems they solve and how they solve them, without going into the mathematics behind the scene.

Analytics in Healthcare: A Practical Introduction is divided into six chapters. Chapter 1 is a brief introduction to data analytics and business intelligence (BI) and their applications in healthcare. Chapter 2 offers a smooth overview of the analytics building blocks with an introduction to basic statistics. Chapter 3 is a detailed explanation of descriptive, predictive, and prescriptive analytics including supervised and unsupervised learning and an example algorithm for each type of analytics. Chapter 4 presents a myriad of applications of analytics in healthcare. Chapter 5 presents health data visualization such as graphs, infographics, and dashboards, with a multitude of visual examples. Chapter 6 delves into the current future directions in healthcare analytics.

Toronto, ON, Canada — Christo El Morr
Toronto, ON, Canada — Hossam Ali-Hassan

Contents

Chapter 1
Healthcare, Data Analytics, and Business Intelligence

Abstract This chapter introduces the healthcare environment and the need for data analytics and business intelligence in healthcare. It overviews the difference between data and information and how both play a major role in decision-making using a set of analytical tools that can be either descriptive and describe events that have happened in the past, diagnostic and provide a diagnosis, predictive and predict events, or prescriptive and prescribe a course of action.

The chapter then details the components of healthcare analytics and how they are used for decision-making improvement using metrics, indicators and dashboards to guide improvement in the quality of care and performance. Business intelligence technology and architecture are then explained with an overview of examples of BI applications in healthcare. The chapter ends with an outline of some software tools that can be used for BI in healthcare, a conclusion, and a list of references.

Keywords Analytics · Business Intelligence (BI) · Data · Information · Healthcare analytics · Metrics · Indicators · BI technology · BI applications

Objectives

By the end of this chapter, you will learn

1. To describe analytics and their use in healthcare
2. To enumerate the different types of analytics
3. To appreciate BI use in healthcare
4. To detail the BI architecture
5. To clearly explain BI and analytics implications in healthcare
6. To give examples of BI applications in healthcare
7. To describe several software tools used for BI

C. El Morr, H. Ali-Hassan, *Analytics in Healthcare*, SpringerBriefs in Health Care Management and Economics, https://doi.org/10.1007/978-3-030-04506-7_1

1.1 Introduction

Today, organizations have access to large amounts of data, whether internal, such as patient/customer detailed profiles and history (medical or purchasing), or external, such as demographics and population data. These data, which are rapidly generated in a very large volume and in different formats, are referred to as big data. In the healthcare field, professionals today have access to vast amounts of data in the form of staff records, electronic patient records, clinical findings, diagnoses, prescription drugs, medical imaging procedures, mobile health, available resources, etc. Managing the data and analyzing it to properly understand it and using it to make well-informed decisions is a challenge for managers and healthcare professionals. Moreover, data analytics tools, also referred to as business analytics or intelligence tools, by large companies such as IBM and SAP and smaller companies such as Tableau and Qlik, are becoming more powerful, more affordable, and easier to use. A new generation of applications, sometimes referred to as end-user analytics or self-serve analytics, are specifically designed for nontechnical users such as business managers and healthcare professionals. The ability to use these increasingly accessible tools with abundant data requires a basic understanding of the core concepts of data, analytics, and interpretation of outcomes that are presented in this book.

What do we mean by analytics? Analytics is the science of analysis—to use data for decision-making [1]. Analytics involves the use of data, analysis, and modeling to arrive at a solution to a problem or to identify new opportunities. Data analytics can answer questions such as (1) what has happened in the past and why, referred to as descriptive analytics; (2) what could happen in the future and with what certainty, referred to as predictive analytics, and (3) what actions can be taken now to control events in the future, referred to as prescriptive analytics [2, 3]. In the healthcare field, analytics can answer questions such as, is there a cancer present in this X-ray image? Or how many nurses do we need during the upcoming holiday season given the patient admission pattern we had last year and the number of patients with flu that we admitted last month? Or how can we optimize the emergency department processes to reduce wait times?

Data analytics have traditionally fallen under the umbrella of a larger concept, called business intelligence, or BI. BI is a conceptual framework for decision support that combines a system architecture, databases and data warehouses, analytical tools, and applications [1]. BI is a mature concept that applies to many fields, including healthcare, despite the presence of the word "business." While remaining a very common term, BI is slowly being replaced by the term analytics, sometimes referring to the same thing. The commonality and differences between BI and analytics will be clarified later in this chapter.

Fig. 1.1 Data to action value chain

1.2 Data and Information

Data are the raw material used to build information; data is simply a collection of facts. Once data are processed, organized, analyzed, and presented in a way that assists in understanding reality and ultimately making a decision, it is called information. Information is ultimately used to make a decision and take a course of action (Fig. 1.1).

1.3 Decision-Making in Healthcare

From an analytics perspective, one can look at healthcare as a domain for decision-making. A nurse or a doctor collects data about a patient (e.g., temperature, blood pressure), reviews an echocardiogram (ECG) screen, and then assesses the situation (i.e., processes the data) and makes a decision on the next step to move the patient forward towards healing. A director of a medical unit in a hospital collects data about the number of inpatients, the number of beds available, the previous year's occupancy in the unit, and the expected flu trends for the season to predict the staffing needs for the Christmas season and make certain decisions about staffing (e.g., vacations, hiring). A radiologist accesses a digital image (e.g., X-ray, ultrasound, computed tomography (CT), magnetic resonance imaging (MRI)), uses the digital image processing tools available on her/his diagnostic workstation to make a diagnosis and reports the presence or absence of a disease. A committee might access admission data, operating room (OR) data, intensive care unit (ICU) data, financial data, or human resource data and use software to prescribe a reorganization of schedules to optimize ED [4, 5], OR [6, 7], and ICU scheduling [8–10].

These are different types of decision-making tasks that require different kinds of analytics that we will explore in detail in Chap. 2. As mentioned above, some of these analytics tools explained above are descriptive of a situation presenting output such as charts and numbers to decision makers, such as the case of the ECG output and the temperature presented to the nurse/doctor. Some other analytics are diagnostic; they present the decision maker with the information necessary to make a diagnosis, such as the case of the software tools used by the radiologist. Some are predictive and assist in making a prediction about the future, such as the case of a software tool used by the director of the medical unit. Finally, other analytics are prescriptive and assist in prescribing a course of action to attain a goal, such as the example of the ED, OR, and ICU scheduling optimization.

1.4 Components of Healthcare Analytics

Data analytics are the systematic access, organization, transformation, extraction, interpretation, and visualization of data using computational power to assist in decision-making. The data are not necessarily voluminous (i.e., big data); there are specific methods for analyzing big data called big data analytics, which are briefly covered in the last chapter of this book.

Trevor Strome's five basic layers of analytics [11] include the following (Fig. 1.2).

1. Business context
2. Data
3. Analytics
4. Quality and performance management
5. Presentation

On the basis of this stack is the business context in which people must define their objectives (including strategic objectives) and measurable goals. In patient-centered care, the voice of the patient is paramount. Once the business context is set and clear, the data context must be defined including the source and quality of the data, its integration, the data management processes, and the infrastructure present or needed to store and manage the data.

The type of analytics is then defined including the tools (e.g., software), the techniques (i.e., algorithms), the stakeholders, the team involved, the data requirements for analysis, the management, and the deployment strategies. The next level consists of defining methods to measure performance and quality, including the processes involved, measurement indicators, achievable targets, and strategies for evaluation and improvement. Finally, the analytics findings are presented in an easy-to-use

Presentation		
Visualization	Dashboards	Reports
Alerts	Mobile	Geospatial
Quality and Performance Management		
Processes	Indicators	Targets
Improvement strategy	Evaluation strategy	
Analytics		
Tools	Techniques	Team
Stakeholders	Requirements	
Deployment	Management	
Data		
Quality	Management	Integration
Infrastructure	Storage	
Business Context		
Objectives	Goals	Patient Voice

Fig. 1.2 Components of healthcare analytics (adapted from Strome [11])

manner to stakeholders/users; hence, visualization options should be explored, including simple reports, graphics-rich dashboards, alerts, geospatial representations, and mobile responsiveness.

1.5 Measurement, Metrics, and Indicators

The amount of data available in hospitals and healthcare organizations is immense. To improve quality and performance, healthcare managers need to make sense of the data available. The objectives are laid out into measurable goals.

For this purpose, managers must set **metrics** [12–16] and **indicators** [17–19]. Metrics are quantitative measurements to measure an aspect of quality or performance in healthcare [11] on a specific scale; on a personal level, blood pressure is a metric that can be used by an individual to measure some aspects of cardiovascular performance/quality. On a system level, hospitals may build many types of metrics to measure their performance and quality of care, for example, the hospital readmission rate within 30 days of discharge, the emergency department wait time, bed occupancy, the length of stay in the hospital, and the number of adverse drug events. An indicator allows managers to detect the state of the current performance and how far it is from a set target.

However, metrics alone are not sufficient; we need to tie a metric to a target goal to determine whether a certain desirable goal has been attained. Metrics that are tied to a certain target (e.g., a certain number target or a range) are called **indicators;** indicators are markers for progress or achievement [20]. Hence, the quality of care and performance of a hospital can be measured by an indicator such as a readmission rate target lower than 7%. If this is justifiable, then any readmission rate above 7% is an indicator of poor quality of care.

Indicators can be consolidated on a screen using different kinds of visualization tools such as figures, charts, colors, or numbers. These indicators displayed in a simple to use and easy to understand way is called a **dashboard**; dashboards display a snapshot of the "health" of an organization (e.g., a hospital). A gradual color scheme is then used to convey the different states of an indicator; for example, a red color usually indicates an "unhealthy" situation (readmission rates considerably above the target), an orange color indicates a situation above the target but not alarming, and a green color indicates situation within the target [21–23]. Examples of dashboards can be seen in Figs. 1.3, 1.4, 1.5, and 1.6.

1.6 BI Technology and Architecture

Laura Madsen defines BI as "the integration of data from disparate source systems to optimize business usage and understanding through a user-friendly interface." [25]. BI is an umbrella term that combines architectures, tools, methodologies,

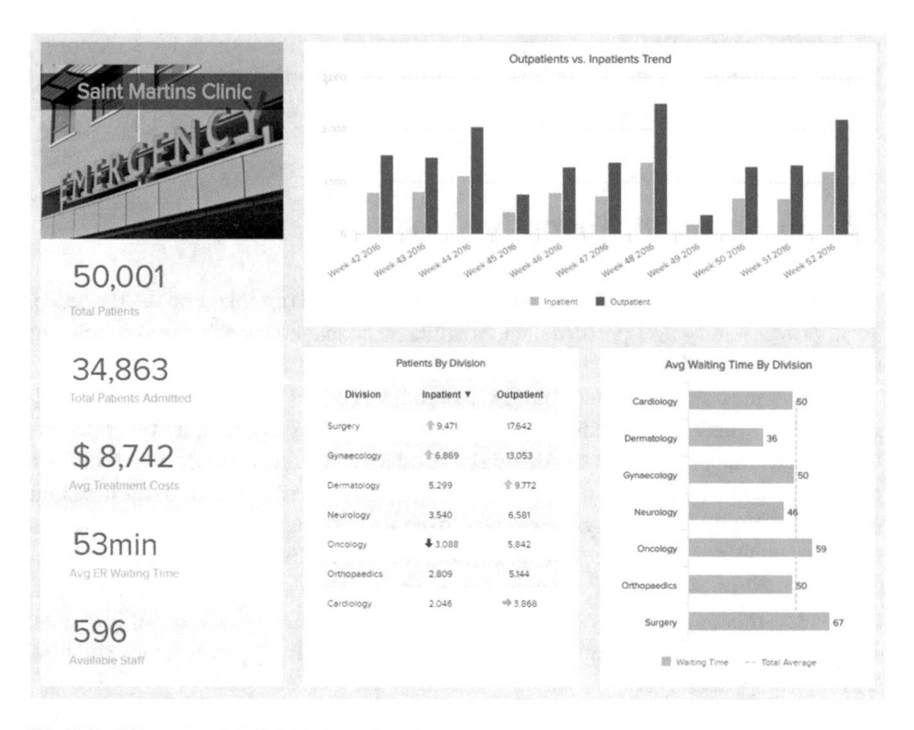

Fig. 1.3 KPI dashboard (Source: datapine.com [24])

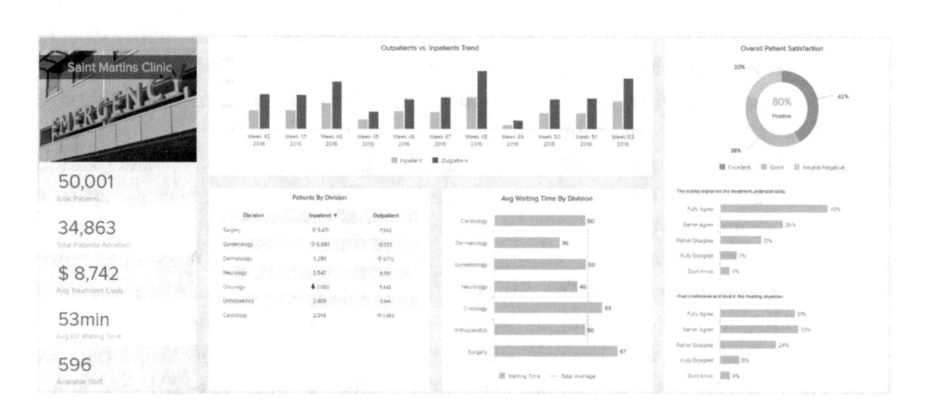

Fig. 1.4 Hospital dashboard (Source: datapine.com [24])

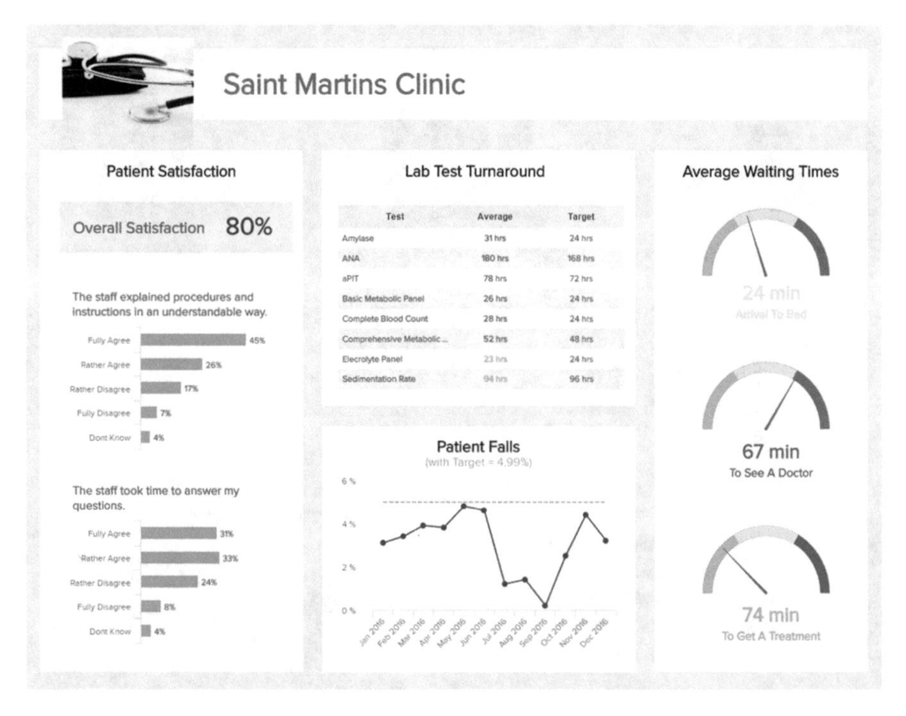

Fig. 1.5 Patient satisfaction dashboard (Source: datapine.com [24])

databases and data warehouses, analytical tools, and applications. The major objective of BI is to enable interactive access to data (and models), to enable manipulation of data and to provide managers, analysts, and professionals with the ability to conduct the appropriate analysis for their needs. BI analyzes historical and current data and transforms it into information and valuable insights (and knowledge), which lead to more informed and better decisions [3]. BI has been very valuable in applications such as customer segmentation in marketing, fraud detection in finance, demand forecasting in manufacturing, and risk factor identification and disease prevention and control in healthcare.

The architecture of BI has four major components: a data warehouse, business analytics, business performance management (BPM), and a user interface. A data warehouse is a type of database that holds source data such as the medical records of patients. It is the cornerstone of medium-to-large BI systems. The data which can be either current or historical are of interest to decision makers and are summarized and structured in a form suitable for analytical activities such as data mining and querying. The second key component is data analytics, which are collections of tools, techniques, and processes for manipulating, mining, and analyzing data stored in the data warehouses. The third key component is business performance management (BPM), which encompasses the tools (business processes, methodologies, metrics, and technologies) used for monitoring, measuring, analyzing, and managing business performance. Finally, BI architecture includes a user interface that

Fig. 1.6 Hospital performance dashboard (Source: datapine.com [24])

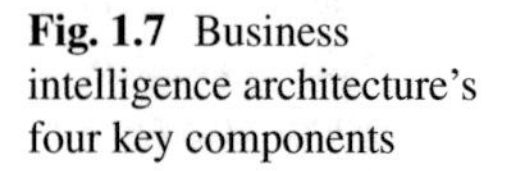

Fig. 1.7 Business intelligence architecture's four key components

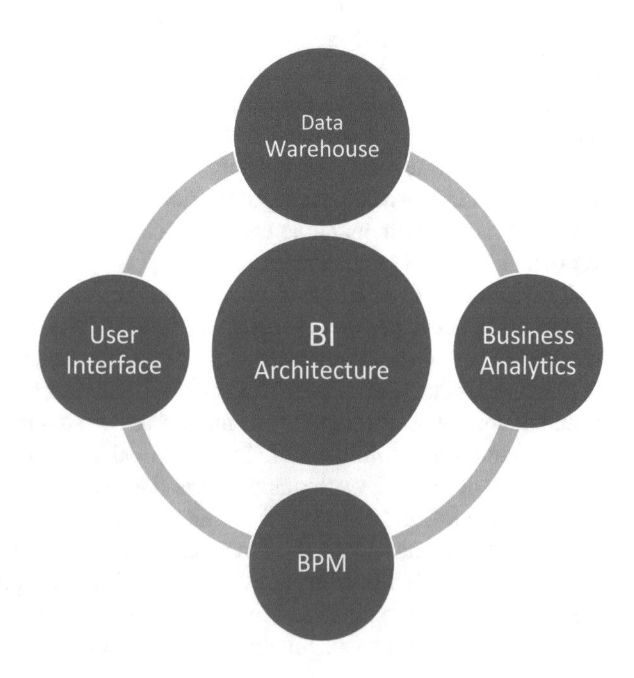

allows bidirectional communication between the system and its user in the form of dashboards, reports, charts, or online forms. It provides a comprehensive graphical view of corporate performance measures, trends, and exceptions [1]. In this book, we will further explore the concepts of data warehouses (Chap. 2), analytics (Chaps. 3 and 4), and user interfaces (Chap. 5) (Fig. 1.7).

1.7 BI Applications in Healthcare

Health organizations need to take actions to be able to measure, monitor, and report on the quality, effectiveness, and value of care. Madsen states that healthcare BI can be defined as "the integration of data from clinical systems, financial systems, and other disparate data sources into a data warehouse that requires a set of validated data to address the concepts of clinical quality, effectiveness of care, and value for business usage" [26]. Data quality, leadership, technology and architecture, and value and culture represent the five facets of healthcare BI (Fig. 1.8).

Examples of BI in healthcare include clinical and business intelligence systems, such as the one implemented at the Broward Regional Health Planning Council in Florida [27], which was built on a regional level to enable healthcare service decision makers, healthcare service planners, and hospitals to access live data generated by many data sources in Florida, including medical facilities utilization data, diagnosis-related group data (DRGs), and health indicator data. The components of such a BI system include extraction, transformation and loading (ETL), a data warehouse, and analytical tools (Fig. 1.9).

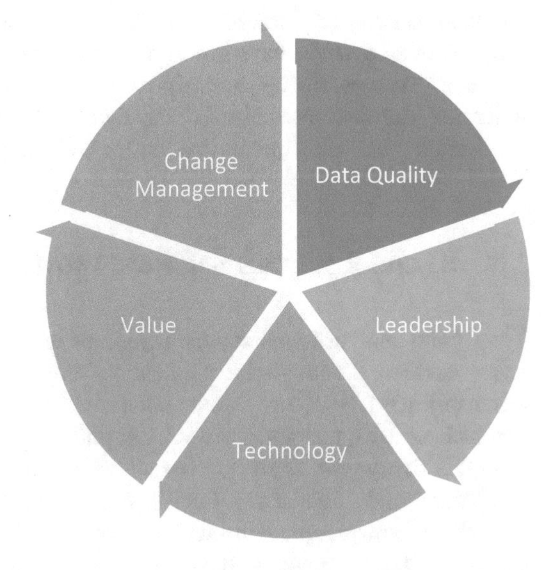

Fig. 1.8 The five facets of healthcare BI (adapted from Laura Madsen's 5 tenets of healthcare BI [26])

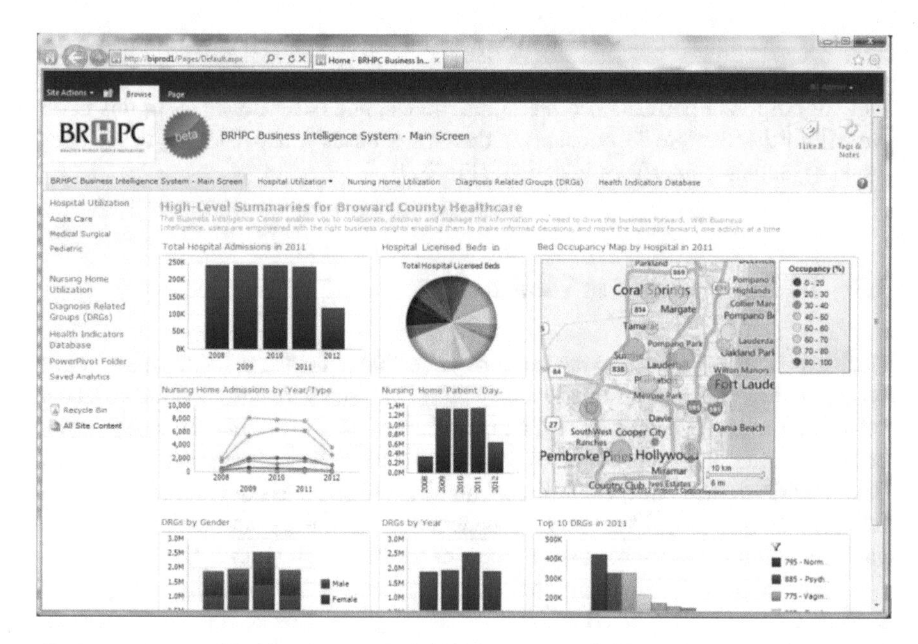

Fig. 1.9 A high-level dashboard of the Broward Regional Health Planning Council business intelligence system (Source: AlHazme et al. [27])

Within radiology, BI can be used to improve quality, safety, efficiency, and cost-effectiveness as well as patient outcomes. The radiology department uses a number of BI metrics [28]; some metrics, such as turnaround time, imaging modality utilization, departmental patient throughput, and wait times, are related to "efficiency"; others relate to quality and safety, such as radiation dose monitoring and reduction and the detection of discrepancies between radiology coding and study reporting [28]. Other BI systems have been proposed to monitor performance by monitoring indicators such as 30-day readmission rates and identifying conditions that most influence readmissions, patients' satisfaction or even monitoring in real-time the medication purchasing and utilization for budgetary/cost purposes [29].

1.8 BI and Analytics Software Providers

The BI and analytics applications landscape is covered by a large number of software vendors. Some of the application providers are software giants such as Microsoft, IBM, SAP, and Oracle, others are large contributors in the field of statistics such as SAS, and some are smaller and specialized providers such as Tableau and Qlik. Every year, Gartner, a consultancy firm, publishes its Magic Quadrant for Analytics and Business Intelligence Platforms (https://www.gartner.com/doc/3861464/magic-quadrant-analytics-business-intelligence). Each year, Gartner places the 20 top vendors in the quadrant based on the completeness of their vision and ability to execute (Fig. 1.10). The companies that score high on both dimensions

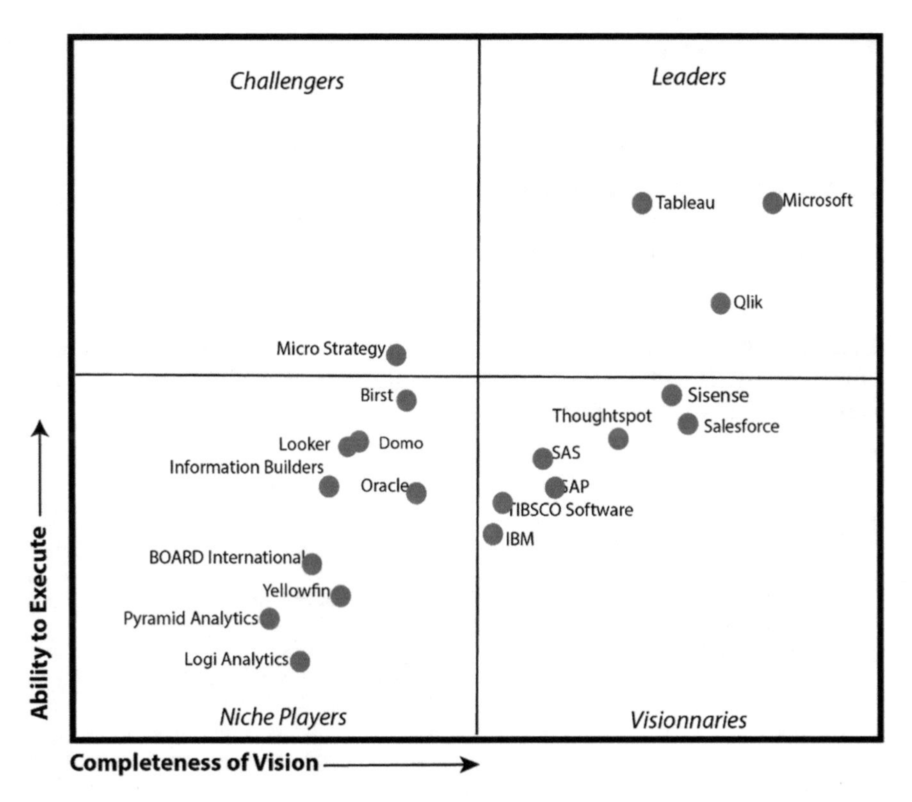

Fig. 1.10 Magic quadrant for analytics and BI platforms (adapted from Gartner Magic quadrant [30])

are labeled as Leaders, and those who score lower are labeled Niche Players. Visionaries are those who score high on completeness but low on the ability to execute while the last quadrant is for Challengers.

In the February 2018 report, three companies led the pack for the third year in a row: Microsoft, Tableau, and Qlik. The next group of vendors that have remained in the quadrant in the past 3 years, moving between Leaders and Visionaries, are SAS, SAP, IBM, and Tibco [31]. The companies listed above are general solution providers for many industries, including healthcare. A recent list of top healthcare business intelligence companies by hospital users was led by Epic Systems, MEDHOST, and Siemens but also included SAS and Qlik [32]. In the Software Toolbox sections of this book, we will focus on providers that are either leaders in the field of analytics or specialize in healthcare analytics.

To obtain a sense of what analytics is and what outcomes it can generate, we suggest you test the different demonstrations provided by Qlik at https://demos.qlik.com/. You can select either of their two products, Qlik Sense or QlikView. The former is focused on the user interface and dashboards, while the latter focuses on analytics. In both cases, you can select the healthcare industry to experience applications such as visualizing operating room management, efficiency and utilization, or analysis of hospital readmissions.

1.9 Conclusion

Paired with abundant data, advanced technology, and easier use, business intelligence (BI) and analytics have recently gained great popularity due to their ability to enhance performance in any industry or field. Analytics, considered by many as part of BI, extracts, manipulates and analyzes data, transforming it into information that helps professionals make well-informed decisions. It supports taking action and generating knowledge. In the healthcare field, analytics plays a major role in areas such as diagnosis, admissions, and prevention. In this chapter, we explored the basic facets of BI with its key components, such as data warehouses and analytical capabilities. Analytics with its four categories, descriptive, diagnostic, predictive, and prescriptive analytics, will be explored in more detail in the next chapter.

References

1. R. Sharda, D. Delen, E. Turban, J. Aronson, and T. P. Liang, *Businesss Intelligence and Analytics: Systems for Decision Support*. 2014.
2. N. Kalé and N. Jones, *Practical Analytics*. Epistemy Press, 2015.
3. R. Sharda, D. Delen, and E. Turban, *Business Intelligence: A Managerial Perspective on Analytics: A Managerial Perspective on Analytics*. Pearson, 2015, pp. 416–416.
4. J. Akin, J. A. Johnson, J. P. Seale, and G. P. Kuperminc, "Using process indicators to optimize service completion of an ED drug and alcohol brief intervention program," (in eng), *Am J Emerg Med,* vol. 33, no. 1, pp. 37–42, Jan 2015.
5. D. Steward, T. F. Glass, and Y. B. Ferrand, "Simulation-Based Design of ED Operations with Care Streams to Optimize Care Delivery and Reduce Length of Stay in the Emergency Department," (in eng), *J Med Syst,* vol. 41, no. 10, p. 162, Sept 6 2017.
6. M. D. Basson, T. W. Butler, and H. Verma, "Predicting patient nonappearance for surgery as a scheduling strategy to optimize operating room utilization in a veterans' administration hospital," (in eng), *Anesthesiology,* vol. 104, no. 4, pp. 826–34, Apr 2006.
7. C. J. Warner *et al.*, "Lean principles optimize on-time vascular surgery operating room starts and decrease resident work hours," (in eng), *J Vasc Surg,* vol. 58, no. 5, pp. 1417–22, Nov 2013.
8. R. Aslakson and P. Spronk, "Tasking the tailor to cut the coat: How to optimize individualized ICU-based palliative care?," (in eng), *Intensive Care Med,* vol. 42, no. 1, pp. 119–21, Jan 2016.
9. J. Kesecioglu, M. M. Schneider, A. W. van der Kooi, and J. Bion, "Structure and function: planning a new ICU to optimize patient care," (in eng), *Curr Opin Crit Care,* vol. 18, no. 6, pp. 688–92, Dec 2012.
10. W. J. Schellekens *et al.*, "Strategies to optimize respiratory muscle function in ICU patients," (in eng), *Crit Care,* vol. 20, no. 1, p. 103, Apr 19 2016.
11. T. L. Strome, *Healthcare analytics for quality and performance improvement*. Hoboken, NJ: Wiley, 2013.
12. C. Daluwatte, J. Vicente, L. Galeotti, L. Johannesen, D. G. Strauss, and C. G. Scully, "A novel ECG detector performance metric and its relationship with missing and false heart rate limit alarms," (in eng), *J Electrocardiol,* vol. 51, no. 1, pp. 68–73, Jan - Feb 2018.
13. K. Honeyford, P. Aylin, and A. Bottle, "Should Emergency Department Attendances be Used With or Instead of Readmission Rates as a Performance Metric?: Comparison of Statistical Properties Using National Data," (in eng), *Med Care,* Mar 29 2018.

14. D. A. Maldonado, A. Roychoudhury, and D. J. Lederer, "A novel patient-centered "intention-to-treat" metric of U.S. lung transplant center performance," (in eng), *Am J Transplant,* vol. 18, no. 1, pp. 226–231, Jan 2018.
15. J. D. Markley, A. L. Pakyz, R. T. Sabo, G. Bearman, S. F. Hohmann, and M. P. Stevens, "Performance of a Novel Antipseudomonal Antibiotic Consumption Metric Among Academic Medical Centers in the United States," (in eng), *Infect Control Hosp Epidemiol,* vol. 39, no. 2, pp. 229–232, Feb 2018.
16. S. Stevanovic and B. Pervan, "A GPS Phase-Locked Loop Performance Metric Based on the Phase Discriminator Output," (in eng), *Sensors (Basel),* vol. 18, no. 1, Jan 19 2018.
17. P. Kaushik, "Physician Burnout: A Leading Indicator of Health Performance and "Head-Down" Mentality in Medical Education-I," (in eng), *Mayo Clin Proc,* vol. 93, no. 4, p. 544, Apr 2018.
18. K. D. Olson, "In Reply-Physician Burnout: A Leading Indicator of Health Performance and "Head-Down" Mentality in Medical Education-I and II," (in eng), *Mayo Clin Proc,* vol. 93, no. 4, pp. 545–547, Apr 2018.
19. J. Peck and O. Viswanath, "Physician Burnout: A Leading Indicator of Health Performance and "Head-Down" Mentality in Medical Education-II," (in eng), *Mayo Clin Proc,* vol. 93, no. 4, pp. 544–545, Apr 2018.
20. Center for Disease Control. (2018, April 22). *Developing Evaluation Indicators*. Available: https://www.cdc.gov/std/Program/pupestd/Developing%20Evaluation%20Indicators.pdf
21. Z. Azadmanjir, M. Torabi, R. Safdari, M. Bayat, and F. Golmahi, "A Map for Clinical Laboratories Management Indicators in the Intelligent Dashboard," (in eng), *Acta Inform Med,* vol. 23, no. 4, pp. 210–4, Aug 2015.
22. M. C. Schall, Jr., L. Cullen, P. Pennathur, H. Chen, K. Burrell, and G. Matthews, "Usability Evaluation and Implementation of a Health Information Technology Dashboard of Evidence-Based Quality Indicators," (in eng), *Comput Inform Nurs,* vol. 35, no. 6, pp. 281–288, Jun 2017.
23. P. Stattin *et al.*, "Dashboard report on performance on select quality indicators to cancer care providers," (in eng), *Scand J Urol,* vol. 50, no. 1, pp. 21–8, 2016.
24. Datapine.com. (2018, September 17). *Healthcare Dashboards Examples*. Available: https://www.datapine.com/dashboard-examples-and-templates/healthcare#patient-satisfaction-dashboard
25. L. Madsen, "Business Intelligence An Introduction," in *Healthcare Business Intelligence: A Guide to Empowering Successful Data Reporting and Analytics*: Wiley, 2012.
26. L. Madsen, "The Tenets of Healthcare BI," in *Healthcare Business Intelligence: A Guide to Empowering Successful Data Reporting and Analytics*: Wiley, 2012.
27. R. H. AlHazme, A. M. Rana, and M. De Lucca, "Development and implementation of a clinical and business intelligence system for the Florida health data warehouse," (in eng), *Online J Public Health Inform,* vol. 6, no. 2, p. e182, 2014.
28. T. S. Cook and P. Nagy, "Business intelligence for the radiologist: making your data work for you," (in eng), *J Am Coll Radiol,* vol. 11, no. 12 Pt B, pp. 1238–40, Dec 2014.
29. B. Pinto and B. I. Fox, "Clinical and Business Intelligence: Why It's Important to Your Pharmacy," (in eng), *Hosp Pharm,* vol. 51, no. 7, p. 604, Jul 2016.
30. C. Howson, R. L. Sallam, J. L. Richardson, J. Tapadinhas, C. J. Doine, and A. Woodward, "Magic Quadrant for Analytics and Business Intelligence Platforms," Gartner Group February 26, 2018. Available: https://b2bsalescafe.files.wordpress.com/2018/03/magic-quadrant-for-analytics-and-business-intelligence-platforms.pdf.
31. B. Aziza. (2018) Gartner Magic Quadrant: Who's Winning In The Data And Machine Learning Space. *Forbes*. Available: https://www.forbes.com/sites/ciocentral/2018/02/28/gartner-magic-quadrant-whos-winning-in-the-data-machine-learning-space/#7dca83b37dab
32. K. Monica, "Top Healthcare Business Intelligence Companies by Hospital Users," ed, 2017.

Chapter 2
Analytics Building Blocks

Abstract This chapter provides an overview of the analytics landscape, including descriptive, diagnostic, predictive, and prescriptive analytics, which are explained in detail with clear examples. A data analytics model that enumerates the steps undertaken during analytics as well as an information management and computing strategy is described.

Keywords Descriptive analytics · Diagnostic analytics · Predictive analytics · Prescriptive analytics · Inferential statistics · Null hypothesis · Correlation · Chi-square · *t*-test · One-way analysis of variance (ANOVA)

Objectives
At the end of this chapter, you will be able to:

1. Compare descriptive, diagnostic, predictive, and prescriptive analytics
2. Describe different statistical tests and their use
3. Appreciate information management and computing strategies

2.1 Introduction

Business intelligence was defined in 1989 as the "the concepts and methods to improve business decision-making by using fact-based support systems" [1]. In the 1990s, new software tools were created to extract, transfer, and load (ETL) large amounts of data in a computer in preparation for analysis. One of the main software tools for BI in that era was Cristal Reports™, currently owned by SAP and marketed for small businesses; it tends to answer questions such as "what happened in a past period of time?," "When?," "Who was involved?," "how many?," and "In what frequency?" As explained in Chap. 1, BI uses a set of metrics to measure *past* performance and report a set of indicators that can guide decision-making; it

C. El Morr, H. Ali-Hassan, *Analytics in Healthcare*, SpringerBriefs in Health Care Management and Economics, https://doi.org/10.1007/978-3-030-04506-7_2

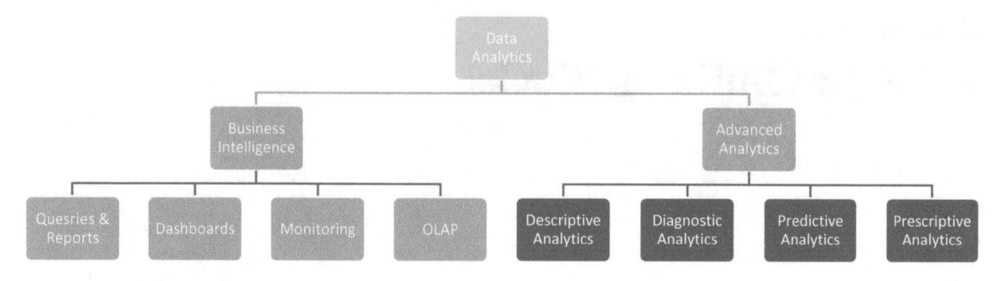

Fig. 2.1 Data analytics types

involves a set of methods such as querying structured data sets and reporting the findings (metrics and key performance indicators), using dashboards, automated monitoring of critical situations (usually involving some threshold). BI is essentially reactive and performed with much human involvement.

Advanced analytics, alternately, are more proactive and performed automatically by a set of algorithms (e.g., data mining and machine learning algorithms). Analytics access structured (e.g., height, weight, and blood pressure) and unstructured data (e.g., free text); they describe "What happened in the past" (Descriptive Analytics), make a diagnosis regarding "Why did it happen?" (Diagnostic Analytics), predict "What will [most likely] happen in the future?" (Predictive Analytics), or even prescribe "What actions should we take to have certain outcomes in the future?" (Prescriptive Analytics). Analytics analyze trends, recognize patterns and possibly prescribe actions for better outcomes, and they use a multitude of methods, such as predictive modeling, data mining, text mining, statistics analysis, simulation, and optimization, which will be covered in the next chapter (Fig. 2.1).

2.2 The Analytics Landscape

2.2.1 Types of Analytics (Descriptive, Diagnostic, Predictive, Prescriptive)

Analytics are of four types: descriptive, diagnostic, predictive, and prescriptive (Fig. 2.2).

2.2.1.1 Descriptive Analytics

Descriptive analytics is another term that is exchangeable with BI, and it queries past or current data and reports on what happened (or is happening). Descriptive analytics displays indicators of past performance to assist in understanding successes and failures and provide evidence for decision-making; for instance, decisions related to delivery of quality care and optimization of performance need to be based on evidence.

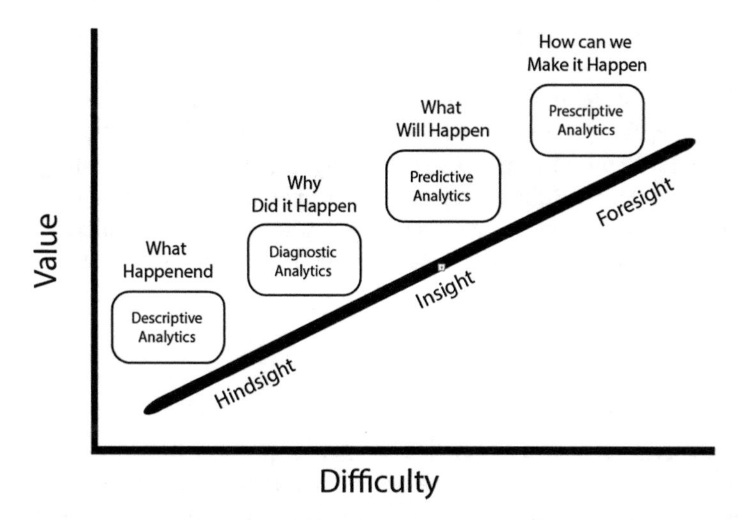

Fig. 2.2 Types of analytics, the value they provide, and their level of difficulty (adapted from Rose Business Technologies [2])

Evidence-based decision-making is of paramount important on the individual health level [3–6], as well as on the managerial and organizational levels [7–10]. Using descriptive analytics, such as reports and data visualization tools (e.g., dashboards), end users can look retrospectively into past events; draw insight across different units, departments and, ultimately, the entire organization; and collect evidence that is useful for an informed decision-making process and evidence-based actions. At the initial stages of analysis, descriptive analytics provide an understanding of patterns in data to find answers to the "What happened" questions, for example, "Who are our patients with recurrent readmission?" and "What are our congestive heart failure patients' ED visits' patterns?" [11]. Descriptive statistics, such as measures of central tendency (mean, median, and mode) and measures of dispersion (minimum, maximum, range, quartiles, and standard deviations), as well as distribution of variables (e.g., histograms), are used in descriptive analytics.

2.2.1.2 Diagnostic Analytics

Descriptive analytics give us insight into the past but do not answer the question "Why did it happen?" Diagnostic analytics aim to answer that type of question. They focus on enhancing processes by identifying why something happened and what the relationships between the event and other variables that could constitute its causes are [12]. They involve trend analysis, root cause analysis [13], cause and effect analysis [14, 15], and cluster analysis [16]. They are exploratory in nature and provide users with interactive data visualization tools [17]. An organization can monitor its performance indicators through diagnostic analysis.

2.2.1.3 Predictive Analytics

A predictive analysis uses past data to create a model that answers the question "What will happen"; it analyzes trends in historical data and identifies what is likely to happen in future. Using predictive analytics, users can prepare plans and implement corrective actions in a proactive manner in advance of the occurrence of an event [17]. Some of the techniques used are what-if analysis, predictive modeling [18–20], machine learning algorithms [21–23], and neural network algorithms [24, 25]. Predictive analytics can be used for forecasting and resource planning.

2.2.1.4 Prescriptive Analytics

While predictive analytics estimate what may happen in the future, prescriptive analytics take a step further by prescribing a certain action plan to address the problems revealed by diagnostic analytics and increase the likelihood of the occurrence of a desired outcome (that may not have been forecasted by predictive analytics) [17, 26–28]. Prescriptive analytics encompass *simulating*, evaluating several *what-if scenarios*, and advising how to maximize the likelihood of the occurrence of desired outcomes. Some of the techniques used in prescriptive analytics are *graph analysis*, *simulation* [29–31], *stochastic optimization* [32–34], and *non-linear programming* [35–37]. Prescriptive analytics are beneficial for advising a course of action to reach a desirable goal. Figure 2.3 provides a snapshot of the evolution of Analytics questions, focus and tools.

2.2.2 Statistics

The basic analytics tools are descriptive statistics, and they are used in BI or descriptive analytics. Readmission rates, the average age of a patient group, and the distribution of patients across Charlson Comorbidity Index values are examples of descriptive statistics.

Other advanced statistical tools are used to infer why an event is happening, and these are called ***inferential statistics***; inferential statistics allow us to draw inferences (i.e., implications) from measurements; they explore relationships between variables, test hypotheses, uncover patterns in data and build predictive models.

In the following, we will overview some common descriptive and inferential statistical tests and their uses. However, we first need to differentiate between the different types of data because data types determine the types of tests we need to use. There are three main types of data: *nominal*, *ordinal*, and *continuous* [39]. Nominal data does not have an established order or rank and has a finite number of values, such as gender and race. Ordinal data has a limited number of options with an implied order, such as number of children. Nominal and ordinal are referred to as discrete data. Continuous data has an infinite number of evenly spaced values, such as blood pressure or height. When collecting any of the three types of data, values

Evolution of Analytics since the 1980s

	DESCRIPTIVE ANALYTICS (FOUNDATIONAL)		DIAGNOSTIC ANALYTICS (OPERATIONAL)	PREDICTIVE ANALYTICS (INSIGNTFUL)	PRESCRIPTIVE ANALYTICS (STRATEGIC)
QUESTION	1- What happened in the past	2- what is happening now?	3-Why did it happen and what are the relationships?	4- What will happen in the future?	5- How would we act in the future?
PROCESS FOCUS	-Reporting	-Measuring KPIs	Trend Analysis Root Cause Analysis Cluster Analysis	Forecasting, Risk Managment, Prediction Probability assessment	Scenario based planning, strategy formulation & simulation, option optimization
TOOLS AND TECHNIQUES	-Static and Interactive Reports	-Dashboard -Performance scoreboards	Data Mining Modeling OLAP, Decision Trees	Data Mining Modeling OLAP, Decision Trees	Liner and non-linear programming, value analysis
	Retrospective			Prospective	

Fig. 2.3 Analytics: questions, focus, and tools (adapted from Podolak [38])

can be grouped into intervals; for example, education level can be categorized into primary school, high school, undergraduate degree, and graduate degree; income might be grouped into categories, such as less than $50,000, [50,000–69,999], [70,000–79,999], and so on, or into low income, medium income, and high income categories. Such data are often referred to as *categorical* data [39].

2.2.2.1 Central Tendency and Dispersion

Central tendency refers to the tendency of scores in a distribution to be concentrated near the middle of the distribution [40]. The most common measures of central tendency for continuous variables are the mean, median, and mode, where each represents a type of average. The most familiar is the arithmetic average, or mean, also known simply as the average. The mean of a set of numbers is the sum of all values that is then divided by the number of observations [41]. Common means in our daily life are the average maximum temperatures in a certain month or the average grade of the students in a course. The mean is particularly useful for summarizing interval or ratio data [40]. The median, which is often confused with the mean, is the value that divides the data such that half of the data points or observations are lower than it and half are higher [41]. A median of 78/100 in an exam means that half the students received a grade below 78 and half received a grade above 78. The median is most useful for summarizing rank order or ordinal scale data but can also be used with interval or ratio scale data [40]. The easiest measure of central tendency to determine is the mode, which is the most common value in a data set [41]. If the mode for an exam is 76/100, it means that the most common grade is 76. Mode is used when we want the quickest estimate of central tendency, when we want to know the largest score obtained by the largest number of subjects, or when we have nominal or categorical data. Median is best used when we have a fairly small distribution with few extreme scores, when the distribution is badly skewed, or when we have missing scores. Finally, the mean is the most useful measure of the three because many statistical tests are based on it and it is more reliable and more stable [40].

Other important statistical measures are measures of variation, such as variance and standard deviation. Deviation represents the distance between each data point, such as patients' blood pressure, and the mean of all observations/measurements. Variance is calculated as the average of the squared deviations of a data set and then summing all the results

$$\text{Variance} = \sum_{i=1}^{n} \left(\text{mean} - \text{observation}_i\right)^2 .$$

The standard deviation is the square root of the variance [41]. Standard deviation can be used to compare the variability in distribution of different data sets and to make a statement about the variability within a data set. For example, a smaller standard deviation means that data values are clustered close to the mean, while a higher standard deviation means that the data points are spread across a larger range [41].

2.2.2.2 Data Distribution

A data distribution is a representation of the spread of the continuous data across a range of values and can be represented by frequency distribution tables, column charts, and histograms. **Distribution** charts can inform us of the level of symmetry or skewness of the data, telling us whether there are roughly as many data points above the mean as there are below it (in the case of symmetry) or whether more observations are above the mean (positive skewness) or below it (negative skewness) [41]. A distribution provides context and helps you better understand your data, such as knowing if a patient's blood pressure is among the highest 5% of all patients.

A special case of data distribution is called the **normal distribution**, also known as the bell curve, which is symmetrical and where the mean, median, and mode are identical; approximately 68% of all the data values lie within one plus or minus standard deviation from the mean, 95% lie within plus or minus two standard deviations from the mean, and nearly all data values lie within plus or minus three standard deviations from the mean [41].

2.2.2.3 Hypothesis Testing, Alpha Levels, Type-I and Type-II Errors

Statistics are often used to test theories or predictions, such as that smoking is associated with lung cancer. In general, this is done by *inference* testing, which is drawing conclusions about a population of interest based on findings from a sample obtained from that population. The specific claim or statement we wish to test, such as "there is a link between smoking and lung cancer," is called a research hypothesis. The first variable, smoking, is called the independent variable, and the second variable, lung cancer status, is called the dependent variable (since we are hypothesizing that its values depend on smoking). The claim that there is **no** link between smoking and lung cancer is called the **null hypothesis** and is denoted as H_0. The alternative hypothesis or research hypothesis is denoted by H_1. When testing the hypothesis (also referred to as statistical inference or significance testing [39]), we assume that the null hypothesis is true, and we try to refute it. If the null hypothesis is rejected after statistical analysis (for example using a t-test or correlation covered later in this chapter), then we can draw a conclusion that the association between lung cancer and smoking is significant.

When we statistically test a hypothesis, we can accept a certain level of significance known as α (alpha). When we say that a finding is "statistically significant," it means that the finding is unlikely to have occurred by chance and that the level of significance is the maximum chance that we are willing to accept [41]. A very common threshold for the level of significance, or α, is 0.05 or 5%, with 0.01 or 1% considered marginally significant [41].

Two types of errors may result from hypothesis testing: Type-I and Type-II errors. Type-I error occurs when we reject the null hypothesis (for example, we conclude that there is a significant association between two variables or that there was a significant difference between the measurements of two or more different group of patients) when in fact the null hypothesis is true (there is no significant association or difference between the variables). Type-II error occurs when we do not reject the null hypothesis when in fact it is false.

If a type-I error is costly, meaning your belief that your theory is correct when it is not could be problematic, then you should choose a low value for α to avoid that error [41].

For example, that the blood pressures of a group that took a new drug are significantly lower than those of a group who took placebo would be considered "costly" (risky for patients), and a low α should be adopted. A common value used for α in this case is 0.01 or 1%. If a type-II error is costly, then you should choose a higher value for α to avoid that error, such as 0.1 or 10%.

2.2.2.4 Statistical Significance and *P* Values

To assess the level of significance of our statistical test (*t*-test, chi-square, correlation, etc.), we depend on an outcome called the *p*-value. A *p*-value is generated by default with different statistical tests. We form a decision rule for our hypothesis testing depending on the *p*-value; if the *p*-value is less than our selected level of significance α, then we cannot accept (i.e., we reject) the null hypothesis [41], and we must conclude that our alternative hypothesis is true. The lower the *p*-value is, the greater the significance of our finding is [41]. The steps followed during hypothesis testing are summarized in Fig. 2.4.

We describe next some of the basic statistical tests for association and difference. Tests of regression will be introduced in Chap. 3.

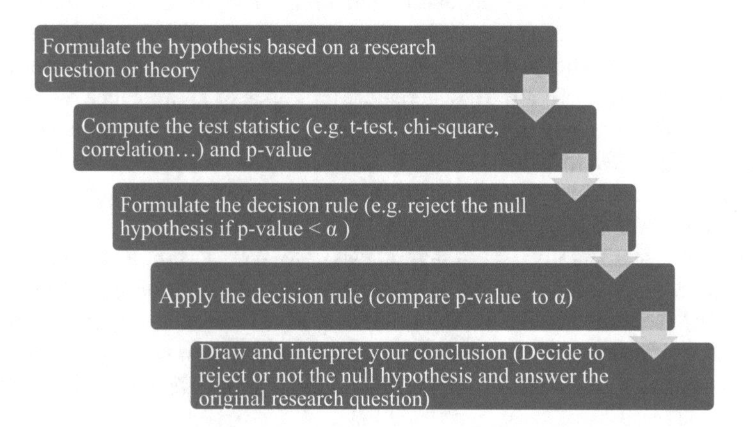

Fig. 2.4 Steps in hypothesis testing (adapted from Nevo [41])

2.2.2.5 Tests of Association

Correlation

Pearson correlation is a test used when both independent and dependent variables have continuous values. In its simplest terms, a linear correlation represents a degree to which a straight line describes the relationship between two variables, such as height and weight. A degree or coefficient of correlation ® ranges from +1 (strong positive correlation or relationship between the two variables) to −1 (strong negative correlation). Values close to zero indicate a weak relationship between the two variables [40, 41]. To test *r* for significance, we propose a null hypothesis, or the assumption that in the population from which our sample was drawn, the two variables, for example, blood pressure and heart disease, are not related [40]. A correlation test using a statistical package, such as MS Excel, SPSS, or SAS, would generate a *p*-value. If the generated *p*-value is less than our selected α, for example, 0.05, then we reject the null hypothesis and conclude that there is an association between blood pressure and heart disease (with a 5% level of significance or risk of type-I error). Figure 2.5 shows an example of correlation test between the age of the patient at admission and the total length of stay at the hospital.

Chi-Square Test of Association

Chi-square is a test used when both independent and dependent variable have categorical values. Chi-square is used to evaluate if there are significant associations between a given exposure (independent variable) and outcome (dependent variable). Commonly, a 2 × 2 table is used to present categorical data where, for example, a column represents exposure or not to a chemical (yes/no) and a row represents a disease or health outcome (yes/no). Each cell represents a count for each category,

Correlations

		Age (at the day of admission)	Length of Stay Total (Length of Stay Acute + Length of Stay ALC)
Age (at the day of admission)	Pearson Correlation	1	.102**
	Sig. (2-tailed)		.000
	N	25389	25389
Length of Stay Total (Length of Stay Acute + Length of Stay ALC)	Pearson Correlation	.102**	1
	Sig. (2-tailed)	.000	
	N	25389	25389

Fig. 2.5 Correlation analysis. "Sig. (2-tailed)" represents the *p*-value, and it is equal to 0.005; the correlation between the jumped distance and the person's height is highly significant

College Majors * Gender Crosstabulation

			Gender		
			Female	Male	Total
College Majors	Humanities	Count	10	4	14
		% within College Majors	71.4%	28.6%	100.0%
		% within Gender	29.4%	17.4%	24.6%
		% of Total	17.5%	7.0%	24.6%
	Natural Sciences	Count	10	11	21
		% within College Majors	47.6%	52.4%	100.0%
		% within Gender	29.4%	47.8%	36.8%
		% of Total	17.5%	19.3%	36.8%
	Social Sciences	Count	14	8	22
		% within College Majors	63.6%	36.4%	100.0%
		% within Gender	41.2%	34.8%	38.6%
		% of Total	24.6%	14.0%	38.6%
Total		Count	34	23	57
		% within College Majors	59.6%	40.4%	100.0%
		% within Gender	100.0%	100.0%	100.0%
		% of Total	59.6%	40.4%	100.0%

Chi-Square Tests

	Value	df	Asymptotic Significance (2-sided)
Pearson Chi-Square	2.215[a]	2	.330
Likelihood Ratio	2.225	2	.329
N of Valid Cases	57		

a. 0 cells (0.0%) have expected count less than 5. The minimum expected count is 5.65.

Fig. 2.6 Chi-square contingency table

and the null hypothesis, for example, can be that that there is no association between a social worker visit and a greater satisfaction with care. We select an α of 5%, for example, and test the data. If the generated p-value is less than the selected α of 0.05, then we reject the null hypothesis and conclude that there is an association between a social worker visit and a greater satisfaction with care (with a 5% level of significance or risk of type-I error) [40] (Fig. 2.6).

2.2.2.6 Test of Difference

Student's *t*-Test

Student's *t*-test is most commonly used to test the difference between the means of the dependent variable (i.e., outcome variable) of two groups, for example, to evaluate if a new anti-hypertensive drug reduces mean systolic blood pressure [39].

Student's t-test is used when one of the variables of interest is continuous (systolic blood pressure) and the other is dichotomous, i.e., nominal with only two values (taking the drug or not). If the new drug is administered to group A of individuals while group B receives a placebo, the null hypothesis would be that there is no difference in the mean systolic blood pressures of the experimental group A and the placebo group B. We select an α of 5%, for example, and test the data, which consists of the systolic blood pressures of the individuals. If the p-value is less than the selected α of 0.05, then we reject the null hypothesis and conclude that there is a difference in the mean between the two groups and that the new drug was effective (with a 5% level of significance or risk of type-I error). This example is known as an independent samples t-test because the two samples are not related. If we test, however, an outcome or dependent variable for the same sample at two different times, or if we match pairs of unrelated individuals (for example, having closely matched behavioral or physiological characteristics that are relevant to the outcome variable), then we call it a dependent or paired-samples t-test [40] (Fig. 2.7).

One-Way Analysis of Variance (ANOVA)

Analysis of Variance, or ANOVA, is similar to the t-test but is used when we want to compare more than two groups at a time. ANOVA is used when one of the variables of interest is continuous and the other is nominal with more than two values, such as three groups A, B, and C. One-way ANOVA examines the effect of one independent variable with comparison to three or more groups, called between-subjects ANOVA, or the same group of subjects at different points of time, called repeated measures ANOVA. For example, to test the effect of a new anti-depression drug, the depression levels of a group of patients are measured before and at several points during the treatment [40] (Fig. 2.8).

2.2.3 Information Processing and Communication

Data processing needs computational power; the needed computer speed depends on the type of analytics used, and it can range from a simple personal computer (PC) using a desktop application, such as SPSS, SAS, or R, to a workstation issuing complex queries to data warehouses, running neural network algorithms, and using data mining tools. Prescriptive analytics might need high performance computing with faster Central Processing Units (CPU), more Random-Access Memory (RAM) and larger and faster storage devices (e.g., hard drives), virtualization capacity (i.e., the ability to allocate large computing capacity to run highly demanding algorithms in terms of computing powers), and the ability to allocate additional capacity on-demand (grid computing and cloud computing) [42]. Networks are used to access data from remote servers holding the data and to communicate results and visualize them to users and stakeholders.

Independent Samples Test

		Levene's Test for Equality of Variances		t-test for Equality of Means						
									95% Confidence Interval of the Difference	
		F	Sig.	t	df	Sig. (2-tailed)	Mean Difference	Std. Error Difference	Lower	Upper
Age (at the day of admission)	Equal variances assumed	5.519	.019	-14.030	25387	.000	-3.180	.227	-3.625	-2.736
	Equal variances not assumed			-14.027	24927.5	.000	-3.180	.227	-3.625	-2.736

Fig. 2.7 *t*-test result in SPSS; the *p*-value shown is 0.019 which means that it is less than 0.05, which reflects a highly significant difference between the two groups tested (Male vs. Female)

Multiple Comparisons

Dependent Variable: Diagnosis based on Beck's Depression Inventory score
Tukey HSD

(I) Medicine administered	(J) Medicine administered	Mean Difference (I-J)	Std. Error	Sig.	95% Confidence Interval Lower Bound	95% Confidence Interval Upper Bound
None	Placebo	.200	.115	.311	-.10	.50
	Homeopathic	.120	.115	.725	-.18	.42
	Pharmaceutical	.680*	.115	.000	.38	.98
Placebo	None	-.200	.115	.311	-.50	.10
	Homeopathic	-.080	.115	.899	-.38	.22
	Pharmaceutical	.480*	.115	.000	.18	.78
Homeopathic	None	-.120	.115	.725	-.42	.18
	Placebo	.080	.115	.899	-.22	.38
	Pharmaceutical	.560*	.115	.000	.26	.86
Pharmaceutical	None	-.680*	.115	.000	-.98	-.38
	Placebo	-.480*	.115	.000	-.78	-.18
	Homeopathic	-.560*	.115	.000	-.86	-.26

*. The mean difference is significant at the 0.05 level.

Fig. 2.8 ANOVA test between three groups of patients, one that was administered a placebo, one a homeopathic drug, and one a pharmaceutical drug. The figure shows that the only group that showed significant improvement in the measurement was the one that had pharmaceutical drugs administered

2.3 Conclusion

Analytics assist us in making decisions by either describing what happened in the past, predicting what might happen in the future, or even prescribing what course of actions ought to be taken to reach a certain goal. This chapter has provided a description of the basic building blocks of the analytics landscape focusing on two key pillars, data and statistics. Databases, data warehouses, and data marts constitute different ways to store and integrate data from many sources, which in turn can be processed, as well as be used for statistical analysis. In this chapter, Descriptive, diagnostic, predictive, and prescriptive analytics were introduced with examples, followed by a data analytics model that enumerates the steps undertaken to execute an analytics project. Because knowledge of basic statistical concepts is necessary for understanding and appreciating the complexity of data analytics, key concepts, such as statistical tests and hypothesis verification, were covered. The following chapter will build upon the material covered thus far and cover descriptive, predictive, and prescriptive analytics in more detail and depth.

References

1. K. D. Lawrence and R. Klimberg, *Contemporary Perspectives in Data Mining, Volume 1.* Information Age Publishing, 2013.
2. RoseBusinessTechnologies.(2013,April26).*DescriptiveDiagnosticPredictivePrescriptiveAnalytics* Available: http://www.rosebt.com/blog/descriptive-diagnostic-predictive-prescriptive-analytics
3. T. Harder *et al.*, "Evidence-based decision-making in infectious diseases epidemiology, prevention and control: matching research questions to study designs and quality appraisal tools," (in eng), *BMC Med Res Methodol,* vol. 14, p. 69, May 21 2014.
4. A. K. Ikeda, P. Hong, S. L. Ishman, S. A. Joe, G. W. Randolph, and J. J. Shin, "Evidence-Based Medicine in Otolaryngology Part 7: Introduction to Shared Decision Making," (in eng), *Otolaryngol Head Neck Surg,* vol. 158, no. 4, pp. 586–593, Apr 2018.
5. A. K. Ikeda, P. Hong, S. L. Ishman, S. A. Joe, G. W. Randolph, and J. J. Shin, "Evidence-Based Medicine in Otolaryngology, Part 8: Shared Decision Making-Impact, Incentives, and Instruments," (in eng), *Otolaryngol Head Neck Surg,* p. 194599818763600, Mar 1 2018.
6. J. A. Spertus, "Understanding How Patients Fare: Insights Into the Health Status Patterns of Patients With Coronary Disease and the Future of Evidence-Based Shared Medical Decision-Making," (in eng), *Circ Cardiovasc Qual Outcomes,* vol. 11, no. 3, p. e004555, Mar 2018.
7. B. M. Niedzwiedzka, "Barriers to evidence-based decision making among Polish healthcare managers," (in eng), *Health Serv Manage Res,* vol. 16, no. 2, pp. 106–15, May 2003.
8. V. Lapaige, "Evidence-based decision-making within the context of globalization: A "Why-What-How" for leaders and managers of health care organizations," (in eng), *Risk Manag Healthc Policy,* vol. 2, pp. 35–46, 2009.
9. E. J. Forrestal, "Foundation of evidence-based decision making for health care managers, part 1: systematic review," (in eng), *Health Care Manag (Frederick),* vol. 33, no. 2, pp. 97–109, Apr-Jun 2014.
10. E. J. Forrestal, "Foundation of evidence-based decision making for health care managers-part II: meta-analysis and applying the evidence," (in eng), *Health Care Manag (Frederick),* vol. 33, no. 3, pp. 230–44, Jul-Sep 2014.
11. H. Geng, *Internet of Things and Data Analytics Handbook.* Wiley, 2017.
12. S. Maloney, "Making Sense of Analytics," presented at the eHealth2018, Toronto ON, Available: http://www.healthcareimc.com/main/making-sense-of-analytics/
13. R. S. Uberoi, U. Gupta, and A. Sibal, "Root Cause Analysis in Healthcare," *Apollo Medicine,* vol. 1, no. 1, pp. 60–63, 2004/09/01/ 2004.
14. W. E. Fassett, "Key performance outcomes of patient safety curricula: root cause analysis, failure mode and effects analysis, and structured communications skills," (in eng), *Am J Pharm Educ,* vol. 75, no. 8, p. 164, Oct 10 2011.
15. R. Ursprung and J. Gray, "Random safety auditing, root cause analysis, failure mode and effects analysis," (in eng), *Clin Perinatol,* vol. 37, no. 1, pp. 141–65, Mar 2010.
16. M. Liao, Y. Li, F. Kianifard, E. Obi, and S. Arcona, "Cluster analysis and its application to healthcare claims data: a study of end-stage renal disease patients who initiated hemodialysis," *BMC Nephrology,* vol. 17, p. 25, 03/02 09/15/received 02/19/accepted 2016.
17. M. Chowdhury, A. Apon, and K. Dey, *Data Analytics for Intelligent Transportation Systems.* Elsevier Science, 2017.
18. H. H. Hijazi, H. L. Harvey, M. S. Alyahya, H. A. Alshraideh, R. M. Al Abdi, and S. K. Parahoo, "The Impact of Applying Quality Management Practices on Patient Centeredness in Jordanian Public Hospitals: Results of Predictive Modeling," (in eng), *Inquiry,* vol. 55, p. 46958018754739, Jan-Dec 2018.
19. F. Noviyanti, Y. Hosotani, S. Koseki, Y. Inatsu, and S. Kawasaki, "Predictive Modeling for the Growth of Salmonella Enteritidis in Chicken Juice by Real-Time Polymerase Chain Reaction," (in eng), *Foodborne Pathog Dis,* Apr 2 2018.
20. M. M. Safaee *et al.*, "Predictive modeling of length of hospital stay following adult spinal deformity correction: Analysis of 653 patients with an accuracy of 75% within 2 days," (in eng), *World Neurosurg,* Apr 17 2018.

21. B. Baessler, M. Mannil, D. Maintz, H. Alkadhi, and R. Manka, "Texture analysis and machine learning of non-contrast T1-weighted MR images in patients with hypertrophic cardiomyopathy-Preliminary results," (in eng), *Eur J Radiol,* vol. 102, pp. 61–67, May 2018.
22. P. Karisani, Z. S. Qin, and E. Agichtein, "Probabilistic and machine learning-based retrieval approaches for biomedical dataset retrieval," (in eng), *Database (Oxford),* vol. 2018, Jan 1 2018.
23. M. R. Schadler, A. Warzybok, and B. Kollmeier, "Objective Prediction of Hearing Aid Benefit Across Listener Groups Using Machine Learning: Speech Recognition Performance With Binaural Noise-Reduction Algorithms," (in eng), *Trends Hear,* vol. 22, p. 2331216518768954, Jan-Dec 2018.
24. Y. Wu, K. Doi, C. E. Metz, N. Asada, and M. L. Giger, "Simulation studies of data classification by artificial neural networks: potential applications in medical imaging and decision making," (in eng), *J Digit Imaging,* vol. 6, no. 2, pp. 117–25, May 1993.
25. J. Zhang, M. Liu, and D. Shen, "Detecting Anatomical Landmarks From Limited Medical Imaging Data Using Two-Stage Task-Oriented Deep Neural Networks," (in eng), *IEEE Trans Image Process,* vol. 26, no. 10, pp. 4753–4764, Oct 2017.
26. E. Chalmers, D. Hill, V. Zhao, and E. Lou, "Prescriptive analytics applied to brace treatment for AIS: a pilot demonstration," (in eng), *Scoliosis,* vol. 10, no. Suppl 2, p. S13, 2015.
27. F. Devriendt, D. Moldovan, and W. Verbeke, "A Literature Survey and Experimental Evaluation of the State-of-the-Art in Uplift Modeling: A Stepping Stone Toward the Development of Prescriptive Analytics," (in eng), *Big Data,* vol. 6, no. 1, pp. 13–41, Mar 2018.
28. S. Van Poucke, M. Thomeer, J. Heath, and M. Vukicevic, "Are Randomized Controlled Trials the (G)old Standard? From Clinical Intelligence to Prescriptive Analytics," (in eng), *J Med Internet Res,* vol. 18, no. 7, p. e185, Jul 6, 2016.
29. G. K. Alexander, S. B. Canclini, J. Fripp, and W. Fripp, "Waterborne Disease Case Investigation: Public Health Nursing Simulation," (in eng), *J Nurs Educ,* vol. 56, no. 1, pp. 39–42, Jan 1, 2017.
30. M. Lee, Y. Chun, and D. A. Griffith, "Error propagation in spatial modeling of public health data: a simulation approach using pediatric blood lead level data for Syracuse, New York," (in eng), *Environ Geochem Health,* vol. 40, no. 2, pp. 667–681, Apr 2018.
31. M. Moessner and S. Bauer, "Maximizing the public health impact of eating disorder services: A simulation study," (in eng), *Int J Eat Disord,* vol. 50, no. 12, pp. 1378–1384, Dec 2017.
32. O. El-Rifai, T. Garaix, V. Augusto, and X. Xie, "A stochastic optimization model for shift scheduling in emergency departments," (in eng), *Health Care Manag Sci,* vol. 18, no. 3, pp. 289–302, Sep 2015.
33. A. Jeremic and E. Khoshrowshahli, "Detecting breast cancer using microwave imaging and stochastic optimization," (in eng), *Conf Proc IEEE Eng Med Biol Soc,* vol. 2015, pp. 89–92, 2015.
34. A. Legrain, M. A. Fortin, N. Lahrichi, and L. M. Rousseau, "Online stochastic optimization of radiotherapy patient scheduling," (in eng), *Health Care Manag Sci,* vol. 18, no. 2, pp. 110–23, Jun 2015.
35. M. A. Christodoulou and C. Kontogeorgou, "Collision avoidance in commercial aircraft Free Flight via neural networks and non-linear programming," (in eng), *Int J Neural Syst,* vol. 18, no. 5, pp. 371–87, Oct 2008.
36. S. I. Saffer, C. E. Mize, U. N. Bhat, and S. A. Szygenda, "Use of non-linear programming and stochastic modeling in the medical evaluation of normal-abnormal liver function," (in eng), *IEEE Trans Biomed Eng,* vol. 23, no. 3, pp. 200–7, May 1976.
37. G. H. Simmons, J. M. Christenson, J. G. Kereiakes, and G. K. Bahr, "A non-linear programming method for optimizing parallel-hole collimator design," (in eng), *Phys Med Biol,* vol. 20, no. 3, pp. 771–88, Sep 1975.
38. I. Podolak, "Making Sense of Analytics," presented at the eHealth 2017, Toronto ON, 2017. Available: http://www.healthcareimc.com/main/making-sense-of-analytics/
39. L. K. Alexander, B. Lopes, K. Ricchetti-Masterson, and K. B. Yeatts. (2018). *Common Statistical Tests and Applications in Epidemiological Literature*.

40. B. M. Thorne and J. M. Giesen, *Statistics for the behavioral sciences*. McGraw-Hill Humanities, Social Sciences & World Languages, 2003.
41. D. Nevo, *Making sense of data through statistics - An introduction*. Legerity Digital Press, 2014.
42. J. Burke, *Health Analytics: Gaining the Insights to Transform Health Care*. Wiley, 2013.

Chapter 3
Descriptive, Predictive, and Prescriptive Analytics

Abstract This chapter provides an overview of the descriptive, predictive, and prescriptive analytics landscape. Data mining is first introduced, followed by coverage of the role of machine learning and artificial intelligence in analytics. Supervised and unsupervised learning are compared, along with the different applications that fall under each. The characteristics and role of reports in descriptive analytics are described, along with the extraction of data in a multidimensional environment. Key algorithms, covering different predictive analytics applications, are described in some detail.

Keywords Data mining · CRISP-DM · Machine learning · Artificial intelligence · Supervised learning · Classification · Regression · Unsupervised learning · Clustering · Dimension reduction · OLAP · Multivariate regression · Multiple logistic regression · Linear discriminant analysis (LDA) · Artificial neural networks (ANNs) · *K*-means · Principal component analysis (PCA)

Objectives
At the end of this chapter, you will be able to:

1. Describe the basics of data mining
2. Understand machine learning and Artificial Intelligence (AI) in analytics
3. Differentiate between supervised and unsupervised learning and their applications
4. Understand how multidimensional data are extracted for reports
5. Understand the different types of algorithms used for predictive analytics
6. Have a general idea about prescriptive analytics

C. El Morr, H. Ali-Hassan, *Analytics in Healthcare*, SpringerBriefs in Health Care Management and Economics, https://doi.org/10.1007/978-3-030-04506-7_3

3.1 Introduction

Knowledge can be modeled in a certain form. If we have well-defined knowledge, we can represent it in an accurate manner, such as a mathematical formula [1]. A model permits us to explain reality, to classify objects, and to predict a value (or if an event will occur) knowing its relationship to other known values. If our knowledge is not complete, then we can approximate reality by learning from previous experiences and predicting an outcome with a certain likelihood of accuracy.

Knowledge can be represented in a computer in the form of mathematical formulae or a certain set of rules stored on a hard disk. Alongside the representation of knowledge, we need to store on a computer a reasoning method, i.e., an algorithm (a series of steps to be followed) to process this knowledge to arrive at an outcome/output (e.g., a decision, classification, or diagnosis).

There are many ways to represent knowledge and process it. Machine learning and data mining are prominent ways to represent and process knowledge and will be introduced in the next paragraphs, and then we will overview the main algorithms for descriptive, predictive, and prescriptive analytics using machine learning algorithms, among other techniques. In the next chapter, we will give examples of healthcare applications of many machine learning algorithms.

3.2 Data Mining

Data mining is a cross-disciplinary field that aims to discover novel and useful patterns within *large* data sets using multiple approaches, including machine learning, statistics, and database systems. The data mining process is automatic or semiautomatic (involves human interaction), and it must lead to patterns that are meaningful to the data stakeholders and provide some advantage (e.g., health or economic) [2].

In data mining in particular, a model exists that provides a framework for project execution; it is the **CRoss Industry Standard Process for Data Mining** (**CRISP-DM**) methodology [3, 4], summarized in the following figure (Fig. 3.1).

In the business understanding phase, the team defines the scope and objectives of the project from the business/stakeholders' point of view and then transforms these objectives into a data mining problem with a defined plan to follow. The data understanding phase is second and entails data collection, exploratory data analysis, and evaluation of the data quality. The data preparation phase follows and involves cleaning the data and preparing it for analysis in later stages, selecting the variables and cases for analysis, and transforming data where needed. The modeling phase follows and comprises selection of modeling techniques and generating and fine-tuning the models. During the evaluation phase, the team evaluates the generated models' quality and effectiveness in achieving the set objectives and makes a final decision on adoption of a model. Finally, the team will deploy the model, which usually involves generating reports.

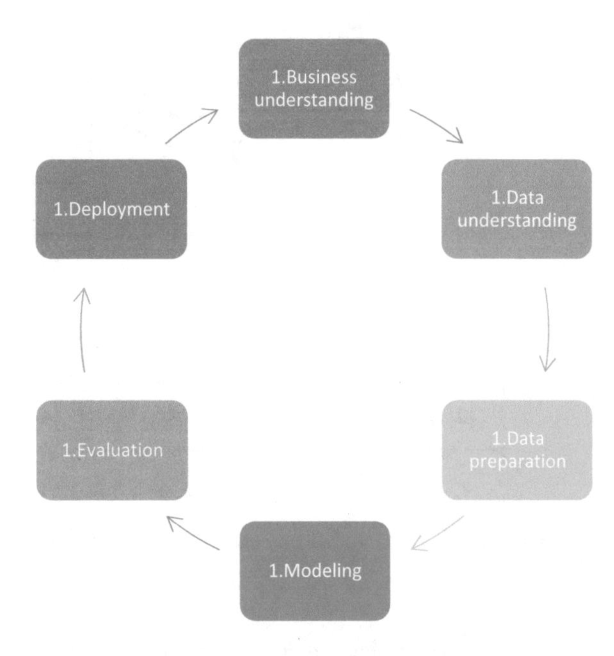

Fig. 3.1 CRISP-DM data mining model

	Prescriptive Analytics (i.e. Business Intelligence)	Predictive Analytics & Prescriptive Analytics
Direction	Past (reactive)	Future (proactive)
Answer questions of the type	What happened? When, how many, who?	What will happen? What's next?
Algorithms/Methods	Key Performance Indicators Metrics Alerts OLAP Dashboards	Predictive Modeling Data Mining Statistics
Data Type	Structured (mostly)	Structured and Unstructured

Fig. 3.2 Comparison between BI and predictive-prescriptive analytics

As we can notice, data mining automates the process of searching for patterns in a large amount of data, and modeling is a core task in data mining. Modeling can be achieved using machine learning methods.

3.3 Machine Learning and AI

While descriptive analytics (i.e., Business Intelligence [5]) are reactive and focus on understanding the past, predictive analytics and prescriptive analytics are proactive and oriented towards the future. Predictive analytics and prescriptive analytics can be defined as the art of constructing models based on historical data and then using them to make predictions [6].

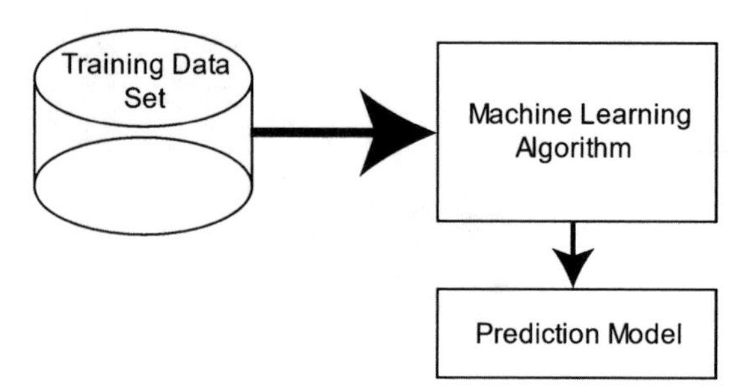

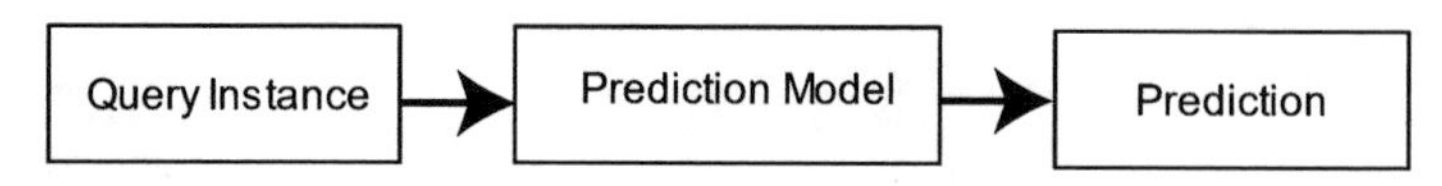

Fig. 3.3 Machine learning: building and using a prediction model (adapted from Kelleher et al. [6])

Instead of answering the "when," "who," and "how many" questions, such as in the BI case, predictive and prescriptive analytics investigate "What will happen" and "what next." Methods in BI relate to *key performance indicators* (**KPI**) or metrics, dashboards, alerts, and OLAP (cubes, slice, dice, and drill), while analytics methods include statistical analysis, predictive modeling, and data mining. BI deals mainly with structured data acted upon by humans, while analytics process structured and unstructured data that are acted upon automatically (or semi-automatically) by computer algorithms (Fig. 3.2).

Artificial Intelligence (**AI**) is a field in computer science dealing with mimicking intelligent behavior; it allows computers to imitate human intelligence, learn from a changing environment, draw conclusions from it and adapt to it. AI uses many types of methods, including a set of methods under the name of "Machine Learning."

Machine learning is an automated process that detects patterns in data [6]; it aims to learn how to improve at tasks with experience and uses many types of techniques, such as Neural Networks and Natural Language Processing (NLP) [7]. **Deep learning** is a subset of machine learning methods that aims to learn from a large amount of data using a set of algorithms, namely, multilayered neural networks.

The idea behind machine learning is that computers can "learn" to accomplish a task by applying a certain algorithm (i.e., a series of steps) on a set of examples (i.e., the **training data set**). Once this training phase is performed, the computers can perform the same task they have learned when they face a new data set that was not part of their learning phase. "Learning" is nothing but building a data model; the training set is the input to the machine learning algorithm, and the model is the output. Subsequently, the model is used to form predictions on new data sets (Fig. 3.3).

Machine learning is performed in either a supervised or unsupervised manner.

3.3.1 Supervised Learning

In **supervised learning**, the training data set contains the input data and the correct output/solution for each data set. The algorithm is presented with a specific data set as well as its solution in advance; it then "learns" how to process these data in a certain way such that it ends up with the provided solution as an output. The learning process is about building a model that ultimately can make predictions of a likely output in the presence of unknown outputs/solutions (i.e., in the presence of uncertainty). Indeed, once learning is performed, the supervised learning software uses its learned model to provide a reasonable output/solution prediction for any new data set input. Supervised learning algorithms can use *classification* and *regression* techniques.

A **classification technique** starts with a set of data and predicts if an output belongs to a certain category/class; for example, for a voice input, it predicts the correct word; for a handwriting image, it predicts the correct word; for a medical image, it predicts a certain diagnosis; for a given patient with known age, weight, height, blood pressure, and other measures, it predicts the likelihood that she/he will have a heart attack within a year.

Some of the main techniques used for classification are:

1. Classification trees
2. Fuzzy classification
3. Random forests
4. Artificial neural networks
5. Discriminant analysis
6. Naive Bayes
7. *K* nearest neighbor
8. Logistic regression (despite its name, it is used in classification)
9. Support vector machine
10. Ensemble methods

Instead of classifying in categories, **a regression technique** predicts a value of an output or variable (i.e., predicts a solution) of a continuous nature (i.e., a number), such as variation in blood pressure, the number of days a patient needs to stay in the hospital, the number of nurses a hospital needed during a holiday season, the number of ED visits for a COPD patient over 60 years of age over the next year, or the cost of a CHF patient to the healthcare system over the next year.

Simply understood, a regression technique approximates a function f of a data input x that produces an output $y = f(x)$; x and y are known, and f is approximated.

Some of the main techniques used in regression are:

1. Linear regression
2. Generalized linear model
3. Decision trees
4. Bayesian networks
5. Fuzzy classification

6. Neural networks
7. Gaussian process regression
8. Relevance vector machine
9. Support vector regression
10. Ensemble methods

A problem such as predicting the number of days the patient will be in good health after discharge from a hospital is a regression problem because we are trying to predict a number. If we are instead interested in predicting if the patient is at high or low risk of readmission to the hospital in the next 30 days, or if we are interested in predicting if the patient will be readmitted or not to the hospital in the next 30 days, then this is a classification problem because we are trying to predict the class that the patient fits in (i.e., low risk vs. high risk, admitted vs. not readmitted) given some of her characteristics (e.g., age, comorbidities).

As you can notice, some techniques are used in both classification and regression.

3.3.2 Unsupervised Learning

In unsupervised learning, there is no known output for the data input, but we have no idea about the possible results we are searching for. The algorithm tries to detect hidden patterns or structures within the data set; once learning is performed, the algorithm will then be able to predict the possible output or solution from a new data set in the future. Two main techniques are used in unsupervised learning: *clustering* and *dimension reduction*.

Clustering aims to find hidden patterns and groupings within the data (i.e., input); it takes a data set as an input and partitions it into clusters. Clustering is helpful in problems such as object recognition, gene sequencing, and market research. Clustering can, for example, identify the group of patients with certain conditions based on healthcare utilization data [8] or healthcare claims data [9].

Some of the main algorithms used in regression are:

1. *K*-means clustering
2. Hierarchical clustering
3. Genetic algorithms
4. Artificial neural networks
5. Hidden Markov model

Dimension reduction is mainly interested in reducing the number of variables/features/attributes (number of dimensions) needed to represent the data input, thus projecting the data set to fewer dimensions. The reduction of the data dimensions (i.e., the number of variables) will allow simpler representation of the data and faster processing time.

Some of the main algorithms used in regression are:

1. Principal component analysis
2. Linear discriminant analysis

3. Multidimensional statistics
4. Random projection

Presented with 100,000 health records for patients with congestive heart failure (CHF) with some readmission history, an algorithm that tries to group patients based on some common characteristics is a clustering algorithm that belongs to the unsupervised learning category. Alternately, if we have 50 characteristics/variables (e.g., age, weight, height, education level, and income level) for each CHF patients, it will be very complex to analyze these characteristics to detect which ones are mostly related to their readmission history, and we can instead use an algorithm, such as Principal Component Analysis (PCA), to detect which (fewer) characteristics most explain the patients' readmission history. Once we have reduced the number of dimensions to a few, 7 for example, we can then proceed with further analysis of the problem.

The following figure gives some examples of the possible applications of supervised and unsupervised learning (Fig. 3.4).

3.3.3 Terminology Used in Machine Learning

The aim of machine learning is to learn or estimate a model of the data set we have and use that model either to predict the class of new data in the future (classification) or the value of an output (regression) or to detect patterns/groups in the data (clustering).

Each instance of the data set is represented by attributes or features; for example, in a particular project dealing with patient readmission, an instance of a patient might be represented by (1) age, (2) gender, (3) length of stay in the hospital, (4) acuity of admission, (5) Charlson comorbidity index, and (6) the number of emergency department visits in the last 6 months prior to the current admission [11]. Each one of these data instances of a particular patient is called a **feature vector**; in the aforementioned patient readmission example, each feature vector in the data set is composed of six features (i.e., each feature vector is of six dimensions); in other words, the **dimensionality** of this data set is six.

The data set we use for learning is called the **training data set**; learning is nothing but the process to generate a model based on the training data. It is important to remember that there is a well-known output for each vector in the training data; suppose we are trying to predict if a patient will be readmitted within 30 days of discharge knowing her/his aforementioned six attributes: our training data should include each patients' readmission attribute (e.g., either yes or no). Once the model is generated (i.e., the learning is performed), then we need to use another data set with known output to validate the model, i.e., to assess the performance of the model and fine-tune its parameters. Such a data set is called the **validation data set**. Once validation is performed, another data set with known output is used to estimate the model error; such a data set is called the **test data set**, and the uncovered error is called the **test error**.

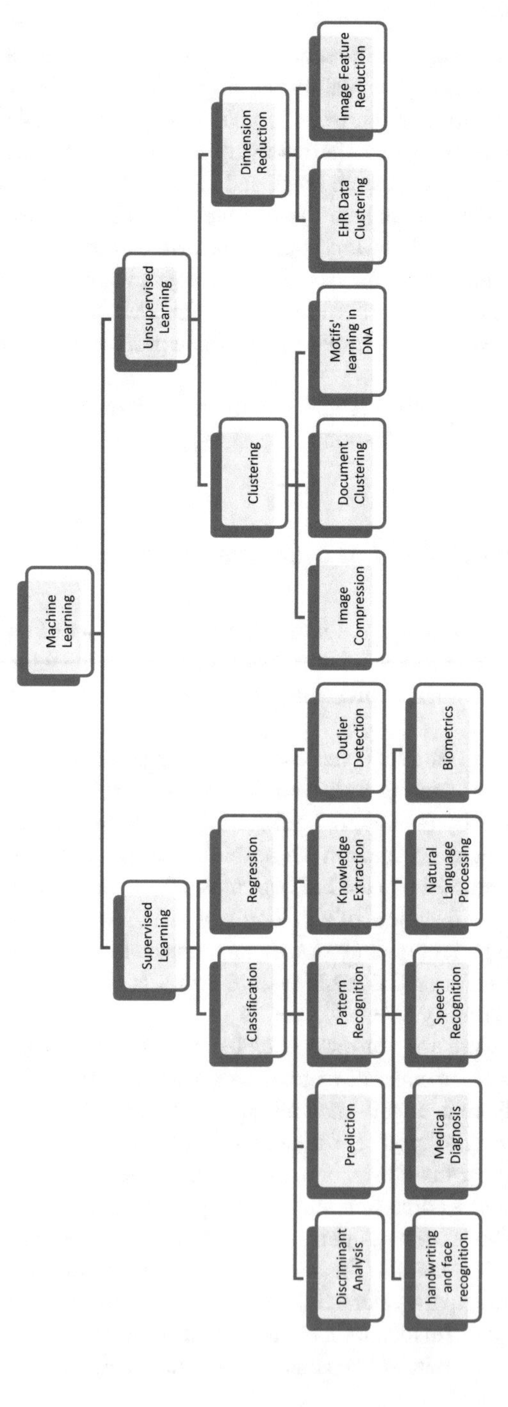

Fig. 3.4 Types of machine learning and examples of applications [10]

A model generated for classification is called a **classifier**; if it is for regression, it is called a **fitted regression model**. Generally speaking, a model that makes predictions is called a **predictor** [12].

3.3.4 Machine Learning Algorithms: A Classification

Machine learning research has been very active in recent years, and it usually deals with large data sets and aims for the creation of statistical models without the need for hypothesis testing. Classifying the techniques into three categories (classification, regression, and clustering and dimensionality reduction) is not simple because some techniques are used across these classes of machine learning styles; however, to organize our understanding of the field, we will provide the following table as a guide, noting that it is not an exhaustive list of techniques and that techniques will appear in more than one category (Figs. 3.5 and 3.6).

3.4 Descriptive Analytics Algorithms

Descriptive analytics have the ability to quantify events and report on them and are a first step in turning data into actionable insights. Descriptive analytics can help with population health management tasks, such as identifying how many patients are living with diabetes, benchmarking outcomes against government expectations, or identifying areas for improvement in clinical quality measures or other aspects of care [14].

3.4.1 Reports

A report is a communication artifact prepared with the specific intention of relaying information in a presentable format [15]. A report is a formal representation of data that has been subject to different analytics techniques, such as slicing, dicing, summarizing, and formatting for the benefit of different users [16]. Reports are an important part of analytics because they are used to support decision-making. They are increasingly becoming visually oriented with colors and graphical icons, looking more like dashboards, with enhanced information content, moving from traditional descriptive reporting to more predictive and prescriptive [15]. Reports can be routinely generated from analytics or can be generated on a one-time basis for a special purpose, referred to as "ad hoc reports." Well-designed reports have fundamental attributes: they should be (1) understandable by the target audience; (2) timely so that they can be useful, especially when the source data changes frequently; (3) accessible, such as on a webpage or mobile device; (4) reliable,

Machine Learning			
	Supervised Learning	Classification	Artificial Neural Networks (ANN) Bayesian Networks (e.g. Naive Bayes) Classification Trees Ensemble Methods Fuzzy Classification K-Nearest Neighbor (KNN) Linear Discriminant Analysis (LDA) Logistic Regression Random Forests Support Vector Machine (SVM)
		Regression	Artificial Neural Networks (ANN) Bayesian Networks (e.g. Naive Bayes) Decision Trees Ensemble Methods Fuzzy Classification Gaussian Process Regression Generalized Linear Model K-Nearest Neighbor (KNN) Linear Regression Multiple Linear Regression Relevance Vector Machine (RVM) Support Vector Regression (SVM)
	Unsupervised Learning	Clustering	Artificial Neural Networks Genetic Algorithms Hidden Markov Model Hierarchical Clustering K-Means Clustering Self Organizing Map
		Dimensionality Reduction	Principal Component Analysis Linear Discriminant Analysis Multidimensional Statistics Random Projection

Fig. 3.5 Machine learning: types of learning and examples of corresponding algorithms; some algorithms can be used in many types of learning

meaning accurate and unbiased and hence trustworthy; (5) complete, including all data that is relevant to a decision; and (6) relevant, leaving out extraneous data that is distracting [16]. Authoring of reports is a multi-step process, starting with the identification of the needs of the users of the report. Then, the sources of the data (TPS systems, functional systems, data warehouses, etc.) are identified, along with the type of analytics that need to be applied to the data to generate the users' information needs. The layout and structure of the report are then defined. The user may be provided with the ability to enter certain pertinent parameters, such as patient identification information and the date of the hospital visit. Finally, reports are deployed in paper or electronic format via emails or websites to their destined users. To be readable, clear, and useful, reports generally contain a combination of the following elements: tables; text; geometric shapes, such as boxes and circle; images; charts; and maps [16]. More details regarding the visualization of data in reports and other mediums can be found in Chap. 5.

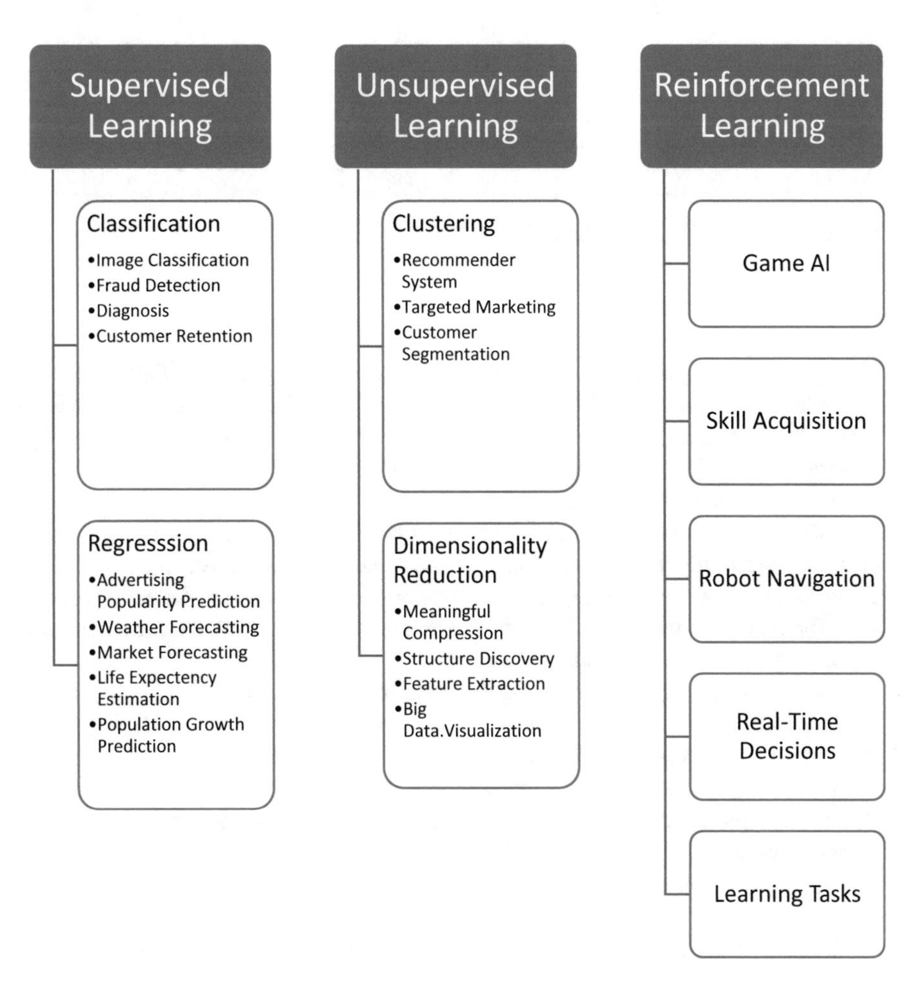

Fig. 3.6 Machine learning applications from a variety of fields (adapted from Vishakha Jha [13])

3.4.2 OLAP and Multidimensional Analysis Techniques

Data to be reported on can be manipulated in many cases with simple arithmetic and statistical operations, such as summing up (e.g., cumulative cost of the treatment of Hepatitis C), counting (e.g., the number of incidents of Hepatitis C), calculating the mean (e.g., the average cost of treating a patient with Hepatitis C), filtering (e.g., extracting names of patients above the age of 75 with Hepatitis C), sorting, ranking, and so on. To extract the data from multiple tables in a relational database, one can issue an SQL Query command that pulls out the data from multiple tables by performing a "join" operation (i.e., joining related data from different tables). To perform Online Analytical Processing (OLAP) on a multidimensional data structure, a number of operations or techniques may be needed, such as slicing, dicing, and pivoting [15–17].

For All Stores and Dates	
Product	Sales in USD
Soda	2,530
Milk	3,800
Juice	15,390
Total	21,720

Fig. 3.7 Slicing (adapted from Ballard et al. [17])

3.4.2.1 Slice and Dice

Slicing and dicing operations are used to make large amounts of data easier to understand and work with. Slicing is a method to filter a large data set into smaller data sets of interest, while dicing these data sets creates even more granularly defined data sets [16]. For simplicity, assume that we have a three-dimensional data set of sales, where the dimensions are: Product, Store, and Date, which can be represented as a cube, where each axis is a dimension and the cells contain sales data, i.e., the product, store, and date of each sale transaction. Slicing is taking a single slice out of the cube, representing one dimension, showing, for example, the sales for each product (Fig. 3.7). In this case, the sales are summed up for all stores and all dates across each product [17].

A dice is a slice on more than two dimensions of the cube [15]. Dicing is putting multiple side-by-side members from a dimension on an axis with multiple related members from a different dimension on another axis, allowing the viewing and analysis of the interrelationship among different dimensions [17]. Two examples of dicing are depicted in Fig. 3.8, showing the sales per store per month and sales per store per product.

3.4.2.2 Pivoting

A **pivot table** is a cross-tabulated structure (crosstab) that displays aggregated and summarized data based on the ways the columns and rows are sorted. Pivoting means swapping the axes or exchanging rows with columns and vice versa or changing the dimensional orientation of a report [15–17] (Fig. 3.9).

3.4.2.3 Drill-Down, Roll-Up, and Drill-across

Drilling down or rolling up is where the user navigates among levels of the data ranging from the most summarized (roll-up) to the most detailed (drill-down) [15] and happens when there is a multi-level hierarchy in the data (e.g., country, province, city, neighborhood) and the users can move from one level to another [17]. Figure 3.10 shows an example of drilling down on the product dimension. When you roll up, the key data, such as sales, is automatically aggregated, and when you drill down, the data are automatically disaggregated [16].

	DATE	1/1/2005	1/2/2005	1/3/2005	Total
	Metrics	Sales in USD	Sales in USD	Sales in USD	Sales in USD
STORE					
CA		40	50	90	**180**
OR		3,115	3,340	1,267	**7,722**
LA		1,583	7,418	4,881	**13,882**
Total		**4,738**	**10,808**	**6,238**	**21,784**

	Product	Milk	Coke	Juice	Total
	Metrics	Sales in USD	Sales in USD	Sales in USD	Sales in USD
STORE					
CA		40	60	80	**180**
OR		60	1,452	6,210	**7,722**
LA		2,430	2,346	9,106	**13,882**
Total		**2,530**	**3,858**	**15,396**	**21,784**

Fig. 3.8 Dicing for store/date (left) and store/product (right) (adapted from Ballard et al. [17])

	Product	Milk	Coke	Juice	Total
	Metrics	Sales in USD	Sales in USD	Sales in USD	Sales in USD
STORE					
CA		40	60	80	**180**
OR		60	1,452	6,210	**7,722**
LA		2,430	2,346	9,106	**13,882**
Total		**2,530**	**3,858**	**15,396**	**21,784**

Pivoting

	Product	Milk	Coke	Juice	Total
	Metrics	Sales in USD	Sales in USD	Sales in USD	Sales in USD
STORE					
CA		40	60	80	**180**
OR		60	1,452	6,210	**7,722**
LA		2,430	2,346	9,106	**13,882**
Total		**2,530**	**3,858**	**15,396**	**21,784**

Fig. 3.9 Pivoting (adapted from Ballard et al. [17])

Drilling across is a method where you drill from one dimension to another, but where the drill-across path must be defined [17]. Figure 3.11 shows an example of a drill-across from the store CA to the product dimension.

3.5 Predictive Analytics Algorithms

The type of algorithm to choose depends on the problem or question at hand. For patient segmentation, for example, one would need clustering algorithms; for a recommender system, the need is for a classification algorithm; to predict the next outcome of time-driven events, one would use a regression algorithm.

3.5.1 Examples of Regression Algorithms

3.5.1.1 Multivariate Regression [12]

Multivariate regression is one of the most common techniques used to create a model that links the dependent outcome variable to the independent variables (e.g., the predictors); it is usually used to produce models for risk predictions in healthcare

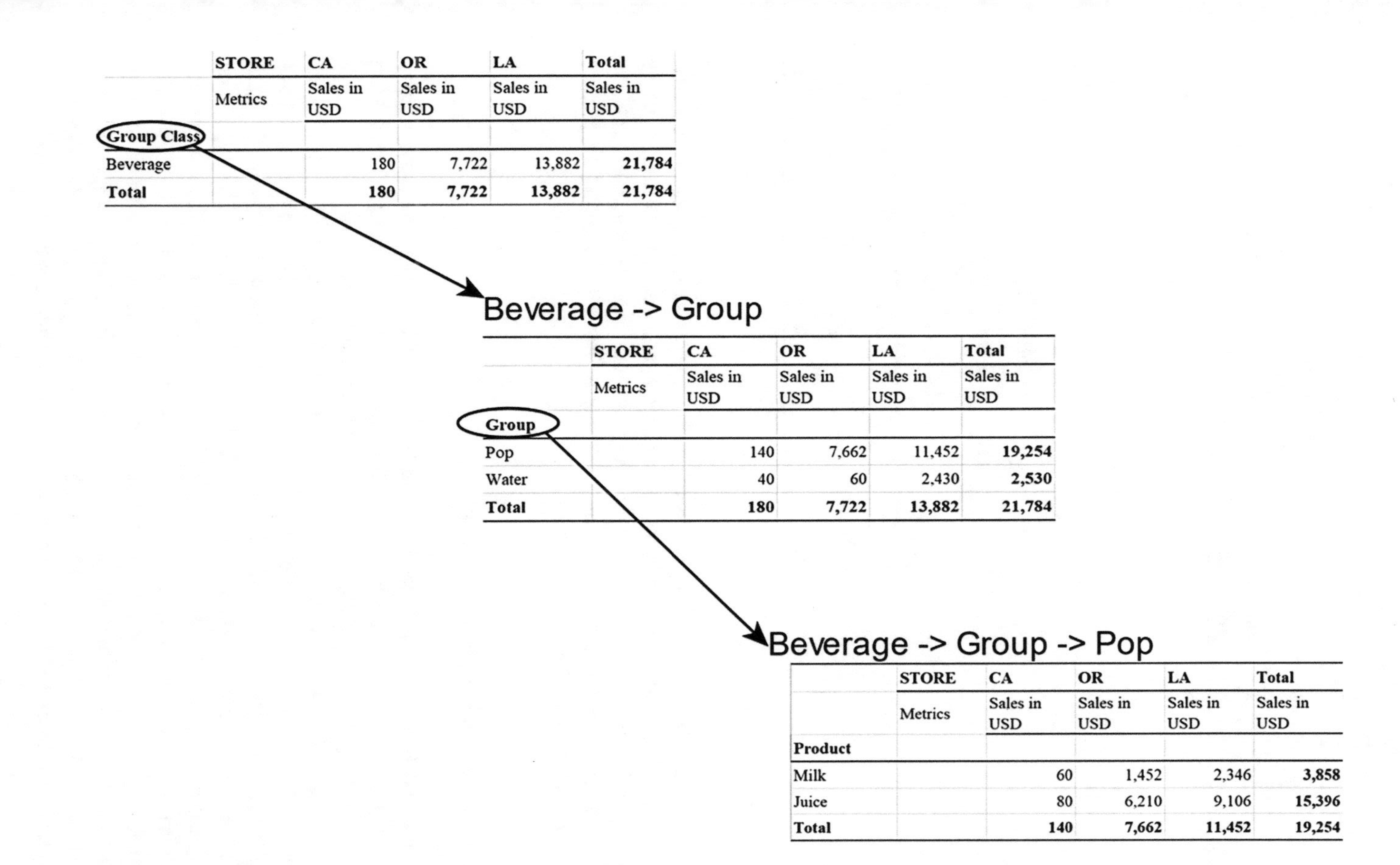

	STORE	CA	OR	LA	Total
	Metrics	Sales in USD	Sales in USD	Sales in USD	Sales in USD
Group Class					
Beverage		180	7,722	13,882	**21,784**
Total		**180**	**7,722**	**13,882**	**21,784**

Beverage -> Group

	STORE	CA	OR	LA	Total
	Metrics	Sales in USD	Sales in USD	Sales in USD	Sales in USD
Group					
Pop		140	7,662	11,452	**19,254**
Water		40	60	2,430	**2,530**
Total		**180**	**7,722**	**13,882**	**21,784**

Beverage -> Group -> Pop

	STORE	CA	OR	LA	Total
	Metrics	Sales in USD	Sales in USD	Sales in USD	Sales in USD
Product					
Milk		60	1,452	2,346	**3,858**
Juice		80	6,210	9,106	**15,396**
Total		**140**	**7,662**	**11,452**	**19,254**

Fig. 3.10 Drilling down (adapted from Ballard et al. [17])

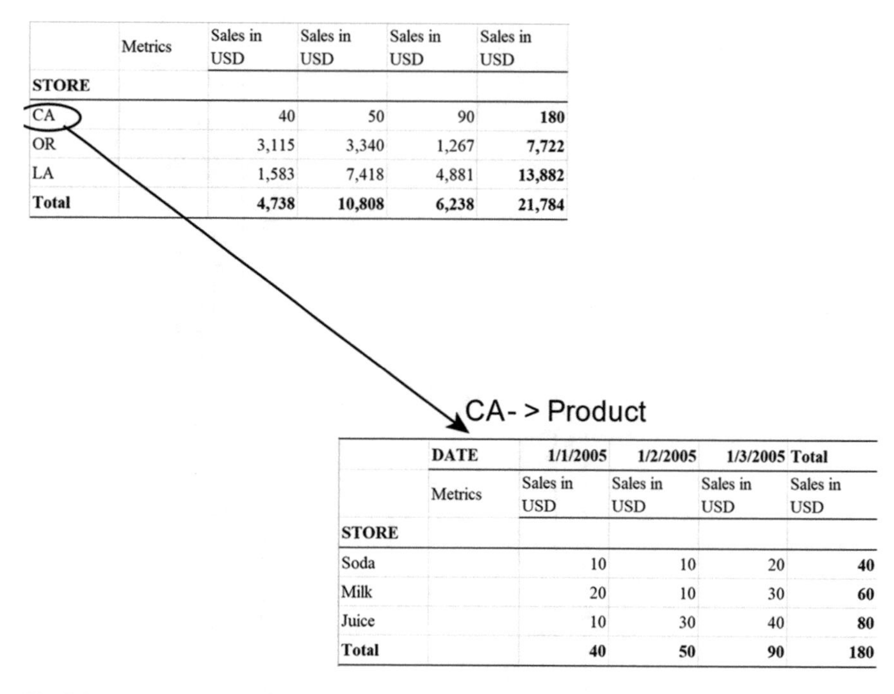

	Metrics	Sales in USD	Sales in USD	Sales in USD	Sales in USD
STORE					
CA		40	50	90	**180**
OR		3,115	3,340	1,267	**7,722**
LA		1,583	7,418	4,881	**13,882**
Total		**4,738**	**10,808**	**6,238**	**21,784**

	DATE	**1/1/2005**	**1/2/2005**	**1/3/2005**	**Total**
	Metrics	Sales in USD	Sales in USD	Sales in USD	Sales in USD
STORE					
Soda		10	10	20	**40**
Milk		20	10	30	**60**
Juice		10	30	40	**80**
Total		**40**	**50**	**90**	**180**

Fig. 3.11 Drilling across (adapted from Ballard et al. [17])

[18]. Multivariate *linear* regression is used when the outcome in question is a continuous variable (i.e., a real number), such as blood pressure, cost, and weight, while multiple *logistic* regression is used when the outcome is categorical, such as blood type, or dichotomous (i.e., binary), such as readmission to the hospital (i.e., yes/no) and risk of readmission (i.e., high/low).

3.5.1.2 Multiple Linear Regressions

In multiple linear regression, the outcome variable (prediction) is expressed in terms of a linear function of the independent variables; for example, healthcare cost = a * (age) + b * (gender) + c * (Carlson Comorbidity Score) + d.

Using past data, a multiple linear regression algorithm can compute the coefficients (a, b, c, d…) for the independent variables, which leads to an expression of their relationship to the dependent example; for instance, cost = 0.5 * (age) + 3 * (gender) + 0.2 * (Carlson Comorbidity Score) + 4 [5]. The mathematical expression represents a *model* that ties the dependent variable (outcome) to the independent variables; the linear logistic regression model can then be used to predict the outcome (e.g., the cost) for any given values for the predictor variables. The score computed by the model can then be a multiplier for the mean (average) outcome variable in the population to predict the outcome for that particular instance/person. In the previous example, for a person whose age is 50 and gender is female (coded as 1) and who has a Charlson comorbidity score of 6, the computed cost score is 0.5 * 50 + 3 * 1 + 0.2 * 6 + 4 = 25 + 3 + 3 + 4 = 35; we would multiply this score (35) by the mean cost in the population for a certain period of time (e.g., a year) to predict the healthcare cost for this person in the future. In the previous example, if the average healthcare cost per year in the population of interest is $2000, then we can predict that the cost for that individual would be 35 * $2000 = $75,000 [18]. Note that, in this example, the higher the predicted score is, the higher the cost is; however, if the predicted score is a positive value that is less than 1, then the cost prediction is lower than the mean in the population.

3.5.1.3 Multiple Logistic Regression

A categorical variable is a variable that can have only a specific number of values; examples of such variables are blood type, gender, and province. Categorical variables with only two possible values are called dichotomous variables.

Logistic regression is used to express a categorical or dichotomous variable as a function of a set of independent variables using one coefficient for each. The model is expressed in a mathematical formula. A main difference between the logistic regression and linear regression is that the predicted value is a *probability* and hence has a value between 0 and 1. The logistic regression model predicts the probability that an observation falls into one of the categories of the dependent variable [5].

Multiple logistic regression is referred to as *polynomial* when it is used to predict a categorical variable, and it is referred to as *binomial* when it is used to predict a dichotomous variable. As in linear regression, a coefficient is created for each predictor variable and used in the regression model to predict individual probabilities of unknown observations' outcomes [18] (Fig. 3.12).

3.5.2 Examples of Classification Algorithms

3.5.2.1 Linear Discriminant Analysis (LDA) [12]

Linear Discriminant analysis (LDA) is a classifier that predicts if an observation falls within a certain class. Given a set of variables, the LDA algorithm separates our observations into different classes by maximizing the distance between the center of these classes and minimizing the distance between the observations within these classes and their center. Hence, LDA reaches a solution where the instances within each class are as close as possible and the instances of the different classes are as far from each other as possible (Fig. 3.13).

A mathematical expression of the LDA algorithm can be written as:

$$\Delta = \frac{\text{maximize}\left(\text{mean } A - \text{mean } B\right)}{\text{minimize}\left(\text{Vairance within groups}\right)}$$

Variables in the Equation

		B	S.E.	Wald	df	Sig.	Exp(B)
Step 1[a]	Com_Com_LACE_High_Low(1)	.974	.124	61.652	1	.000	2.648
	Constant	-2.133	.056	1468.004	1	.000	.119

a. Variable(s) entered on step 1: Com_Com_LACE_High_Low.

Fig. 3.12 Example from SPSS

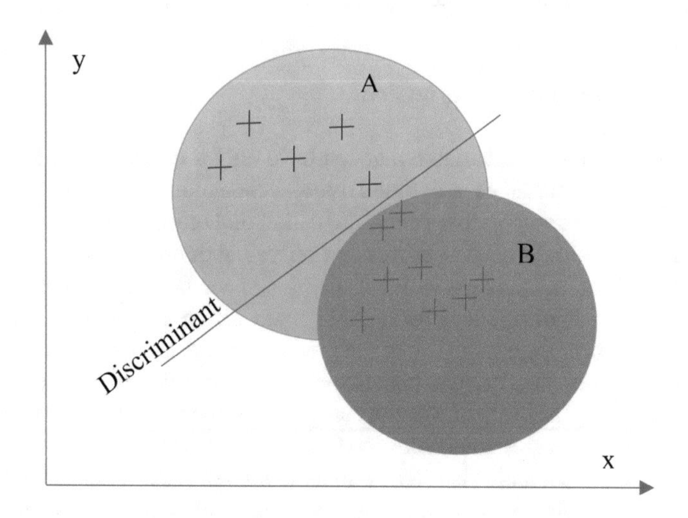

Fig. 3.13 Separation of two classes A and B by one discriminant using LDA

Note that there might be more than two classes; the following figure is an example of three classes separated during LDA (Fig. 3.14).

3.5.2.2 Artificial Neural Networks

Artificial Neural Networks (ANNs) are a computer technique that models the functioning of the brain's neurons; the ANN software can be trained to "learn" how to recognize patterns and classify data [12].

A neuron is a software element that uses an **activation function** (*f*), a set of *adaptive weights* ($w_0 \ldots w_n$), and data *input* ($x_0 \ldots x_n$) to generate an *output*

$$y = f\left(\sum_{i=1}^{n}\left(w_i^{*} x_i\right)\right) = f\left(w_0^{*} x_0 + w_1^{*} x_1 + w_1^{*} x_1 \ldots\right)$$

.

The input vector ($x_0, x_1, x_2 \ldots$) can be sent from another neuron or from other data sources (e.g., observations). The weights are the parameters of the data model (Fig. 3.15).

The artificial neural network is composed of one input layer of neurons, one or more hidden layers, and one output layer [5] (Fig. 3.16).

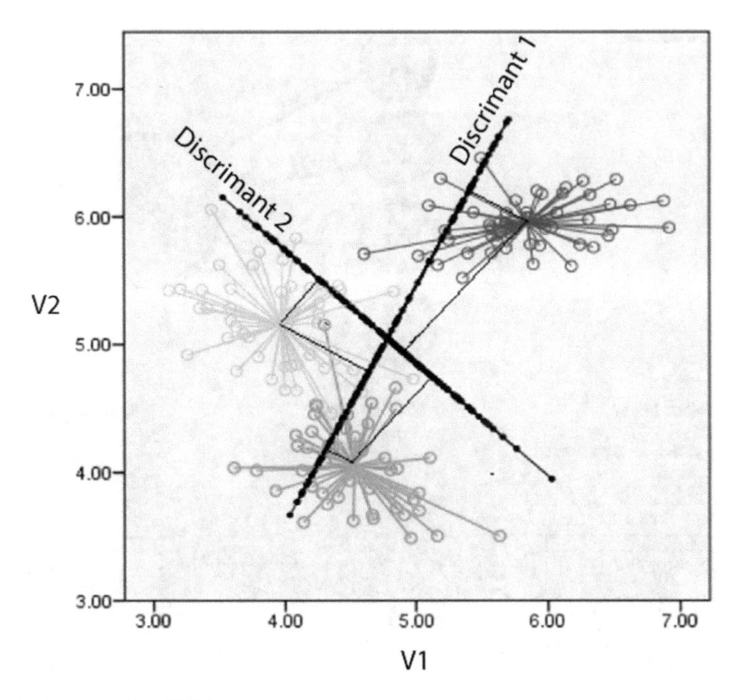

Fig. 3.14 An example of LDA

The aim of the neural network learning process (supervised and unsupervised) is to adjust the model's *weights* to arrive at the correct output, i.e., choose the correct class, arrive at the correct value, or recognize the correct patterns. In a supervised learning environment, the adjustment of the weights is performed by comparing the ANN output to a known output class for the input used; if the output is not correct, then the weights are adjusted iteratively until a correct classification is found (Fig. 3.17).

There are many types of ANNs; one of the most commonly used ones is the Multi-Layer Perceptron (MLP).

3.5.3 Examples of Clustering Algorithms

3.5.3.1 *K*-Means

K-means is a common algorithm used for cluster analysis. Cluster analysis is an essential data mining method to classify items, concepts, or events into common groupings called clusters. *K*-means is an exploratory data analysis tool for solving classification problems by sorting cases into clusters or groups so that the degree of association in a cluster is strong among its members and weak with members of other clusters [20]. *K*-means is a common method in multiple disciplines, such as

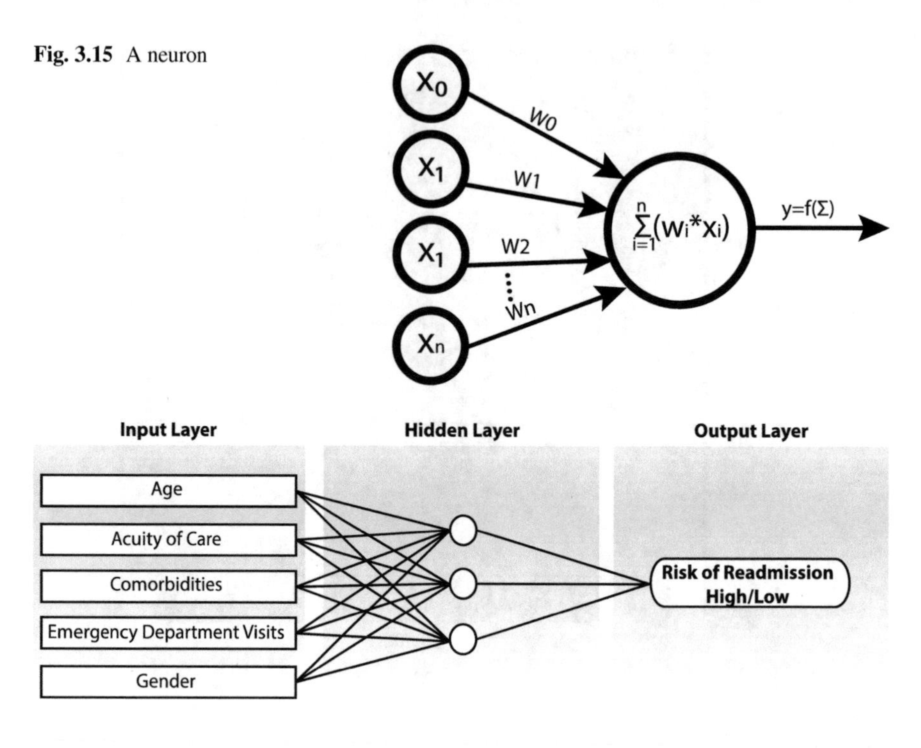

Fig. 3.15 A neuron

Fig. 3.16 An artificial neural network with an input layer with five nodes each for one variable of the input data, one hidden layer, and one node in the output layer representing two classes: high risk of hospital readmission and low risk of hospital readmission (adapted from El Morr [19])

businesses with applications, including market segmentation (classification) of customers and fraud detection [20]. In medicine, *K*-means has been used, for example, for the classification of healthcare claims of end-stage renal disease patients [9] and for examining the heterogeneity of a complex geriatric population [21]. *K*-means, where *K* represents a predetermined number of clusters and where each input belongs to the cluster with the nearest mean, is one of the most referenced clustering algorithms and is one of the best known predictive analysis algorithms [20, 22].

K-means starts by selecting the optimal number of clusters *K*. The selection can be performed using multiple methods; one of the simplest is to compute $K = (n/2)^{1/2}$, where n is the number of data points (i.e., observations). The statistical software used to apply *K*-means generates *K* random points as cluster centers, and each data point or record is assigned to the nearest cluster center. The mean of the data assigned for each cluster is computed and considered the new cluster center, and then the last two steps (assignment of points to the centers and computation of new centers) are repeated until convergence is achieved or the optimal centers are identified [20]. The centers are chosen to minimize the sum of the squares of the distances between a data point and the center point of its cluster and minimize the total intra-cluster variance [16]. *K*-means is simple and logical and thus is widely accepted for cluster analysis [22] (Fig. 3.18).

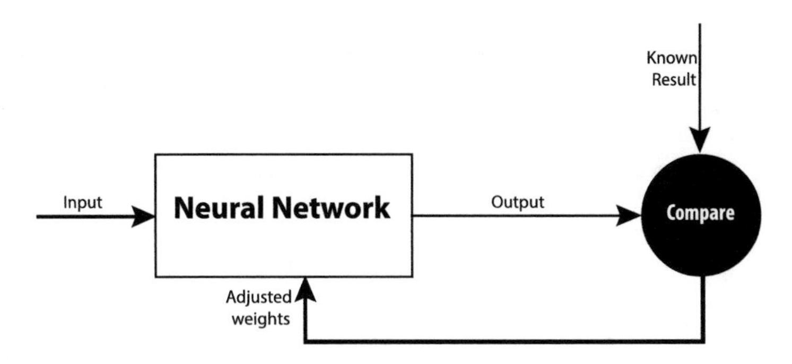

Fig. 3.17 Overall functioning of artificial neural network supervised learning

3.5.4 Examples of Dimensionality Reduction Algorithms

3.5.4.1 Principal Component Analysis (PCA)

A data dimension is a characteristic, feature, or attribute of the data that helps us understand it better and answer basic questions about it. Examples of dimensions are time, location, blood pressure, and age. Dimension reduction is particularly necessary when we have large data sets that have many features or dimensions that make any interpretation of the data, or its use for machine learning, almost impossible or very cumbersome. To address the problem, we look for the dimensions that have the highest variance, where the data varies the most and hence has a high power of explanation [23].

The objective of Principle Component Analysis (PCA), one of the most popular dimensionality reduction methods currently used, is to find the principle components of the data by identifying a hyperplane that lies closest to the data points and projecting the data onto it [23, 24]. PCA hence reduces the data down into its basic components, stripping away any unnecessary parts [23]. To clarify this, we display data in a two-dimensional space (Fig. 3.19). We then look for the two straight lines that are orthogonal and that represent the largest variance or distribution of the data. The lines are referred to as **Eigenvectors**, which come with calculated Eigenvalues that represent how much the data is spread out. The Eigenvector, or line with the highest Eigenvalue, is hence the principle component [23]. In reality, multiple Eigenvectors are identified in a certain data set, but only the ones with an Eigenvalue higher than a certain *threshold* are retained, and the ones with an Eigenvalue lower than the threshold are ignored due to the low variance of the data and hence low explanation value.

A common example of PCA comes from the Oxford Internet Survey (OxIS) 2013 study on Internet use in Britain. The survey asked 2000 people approximately 50 questions that could uncover varying factors of their Internet use, such as age, income, profession, and so on. Trying to analyze 50 potential variables or dimensions to understand Internet use would be very complex and require a large amount of time, effort, and money. PCA was used to reduce the data to its principal components, greatly

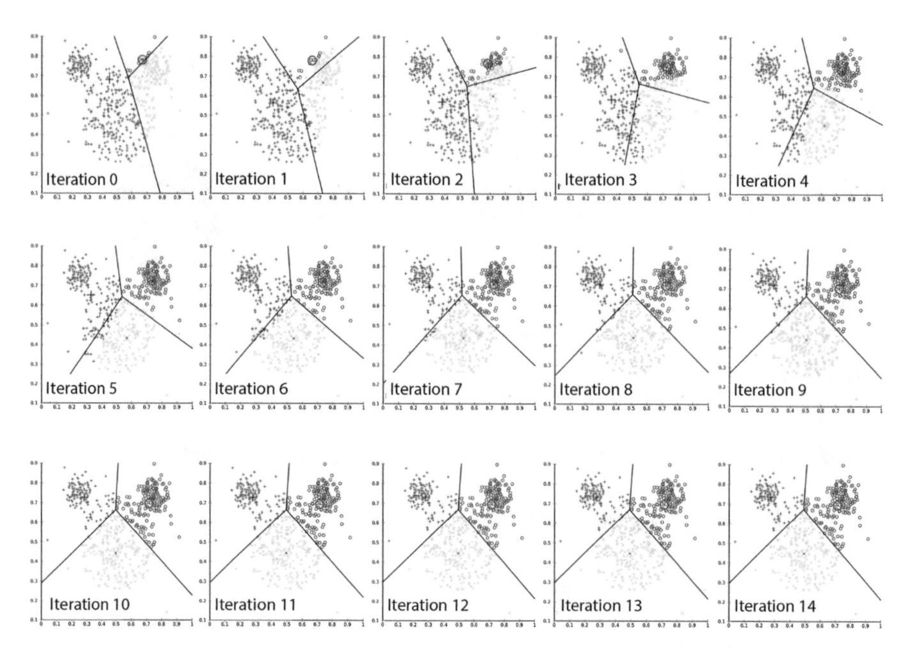

Fig. 3.18 Example of *K*-means convergence (By Chire CC BY-SA 4.0 (https://creativecommons.org/licenses/by-sa/4.0), from Wikimedia Commons)

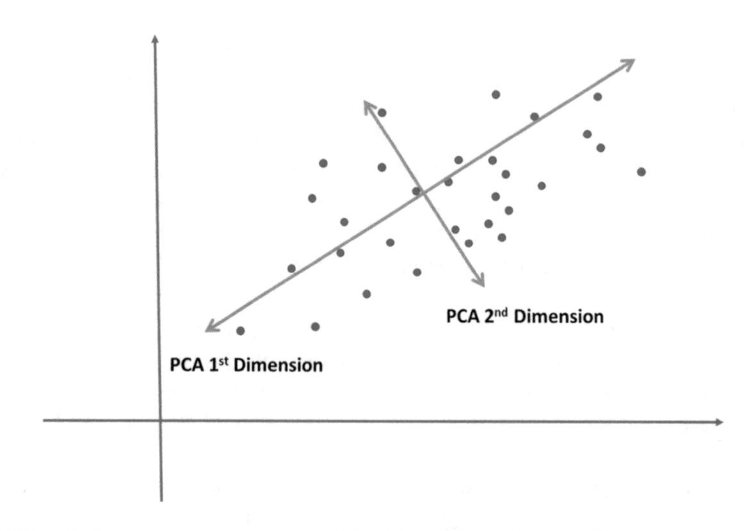

Fig. 3.19 Principal component analysis (PCA) (adapted from Chung [24])

reducing the number of variables that needed to be analyzed from 50 down to 4, which were named: Enjoyable Escape, Instrumental Efficiency, Social Facilitator, and Problem Generator. These dimensions represent the attitudes and beliefs that captured the most variance in the data with Eigenvalues greater than 1 and hence were most valuable. Thus, PCA eliminated factors that have insignificant values instead of attempting to analyze each and every component within the survey [25].

PCA was also applied in a clinical study to investigate the value of early systemic inflammation in predicting ICU-acquired weakness. Systemic inflammation can be represented by a large number of inflammatory cytokines, where each type of cytokine is regarded as one dimension, resulting in dozens of dimensions in the data. PCA was applied, and most of the variance was accounted for by only three principle components; hence, the prediction model was significantly simplified [26, 27].

3.6 Prescriptive Analytics

Prescriptive analytics models add to predictive analytics the ability not only to predict but also to explain why an event happened through a set of rules that are easy to interpret, which allows us to act based on the event using those rules. Some of the predictive analytics algorithms, such as ANNs, do not allow us to understand why a prediction was made; others will and thus allow us to create rules that are actionable (immediately usable), such as Decision trees, Fuzzy Rule-Based Systems, Switching Neural Networks (SNNs), and Logic Learning Machines (LLMs), which are a well-known efficient implementation of SSNs. These algorithms will not be covered in this introductory book.

3.7 Conclusion

This chapter attempted to simplify key analytics concepts while providing enough explanation to understand them and appreciate their complexity, value, and impact. Data mining was first introduced, followed by coverage of the roles of machine learning and artificial intelligence in analytics. Supervised and unsupervised learning were compared, along with the different applications that fall under each. The characteristics and role of reports in descriptive analytics were described, along with the extraction of data in a multidimensional environment. Key algorithms, covering different predictive analytics applications, were described in some detail. Machine learning algorithms can be supervised or unsupervised; supervised learning uses classification and regression techniques, while unsupervised learning uses clustering and dimension reduction. The next chapter will build upon the knowledge attained here by providing examples of healthcare applications of the three categories of analytics (descriptive, predictive, and prescriptive), with a focus on some key algorithms.

References

1. E. Coiera, "Computational Reasoning Methods," in *Guide to Health Informatics, Third Edition*: CRC Press, 2015.
2. I. H. Witten, E. Frank, M. A. Hall, and C. J. Pal, *Data Mining: Practical Machine Learning Tools and Techniques*. Elsevier Science, 2016.
3. Smart Vision Europe. (2018, April 27). *What is the CRISP-DM methodology?* Available: https://www.sv-europe.com/crisp-dm-methodology/
4. D. T. Larose and C. D. Larose, *Discovering Knowledge in Data: An Introduction to Data Mining*. Wiley, 2014.
5. J. Fahl, *Data Analytics: A Practical Guide to Data Analytics for Business, Beginner to Expert*. CreateSpace Independent Publishing Platform, 2017.
6. J. D. Kelleher, B. M. Namee, and A. D'Arcy, *Fundamentals of Machine Learning for Predictive Data Analytics: Algorithms, Worked Examples, and Case Studies*. MIT Press, 2015.
7. B. Marr. (2016, April 30). *What Is The Difference Between Artificial Intelligence And Machine Learning?*
8. T. Lefèvre, C. Rondet, I. Parizot, and P. Chauvin, "Applying Multivariate Clustering Techniques to Health Data: The 4 Types of Healthcare Utilization in the Paris Metropolitan Area," *PLOS ONE,* vol. 9, no. 12, p. e115064, 2014.
9. M. Liao, Y. Li, F. Kianifard, E. Obi, and S. Arcona, "Cluster analysis and its application to healthcare claims data: a study of end-stage renal disease patients who initiated hemodialysis," *BMC Nephrology,* vol. 17, p. 25, 2016.
10. E. Alpaydin, *Introduction to Machine Learning*. MIT Press, 2014.
11. C. El Morr, L. Ginsburg, S. Nam, and S. Woollard, "Assessing the Performance of a Modified LACE Index (LACE-rt) to Predict Unplanned Readmission After Discharge in a Community Teaching Hospital," *Interact J Med Res,* vol. 6, no. 1, p. e2, 03/08 2017.
12. Z. H. Zhou, "Introduction," in *Ensemble Methods: Foundations and Algorithms*(Chapman & Hall/CRC machine learning & pattern recognition series: CRC Press, 2012.
13. V. Jha. (2017, April 30). *Machine Learning Algorithm - Backbone of emerging technologies*. Available: https://www.techleer.com/articles/203-machine-learning-algorithm-backbone-of-emerging-technologies/
14. J. Bresnick. (2015). *Healthcare Big Data Analytics: From Description to Prescription*. Available: https://healthitanalytics.com/news/healthcare-big-data-analytics-from-description-to-prescription
15. R. Sharda, D. Delen, and E. Turban, *Business intelligence: a managerial perspective on analytics*. Prentice Hall Press, 2014.
16. N. Kalé and N. Jones, *Practical Analytics*. Epistemy Press, 2015.
17. C. Ballard, D. M. Farrell, A. Gupta, C. Mazuela, and S. Vohnik, *Dimensional Modeling: In a Business Intelligence Environment*. IBM Redbooks, 2012.
18. M. L. Sylvia and M. F. Terhaar, *Clinical Analytics and Data Management for the DNP, Second Edition*. Springer Publishing Company, 2018.
19. C. El Morr, *Introduction to Health Informatics: A Canadian Perspective*. Canadian Scholars' Press, 2018.
20. R. Sharda, D. Delen, E. Turban, J. Aronson, and T. P. Liang, *Businesss Intelligence and Analytics: Systems for Decision Support*. 2014.
21. J. J. Armstrong, M. Zhu, J. P. Hirdes, and P. Stolee, "K-means cluster analysis of rehabilitation service users in the home health care system of Ontario: Examining the heterogeneity of a complex geriatric population," *Archives of physical medicine and rehabilitation,* vol. 93, no. 12, pp. 2198–2205, 2012.
22. J. MacGregor, *Predictive Analysis with SAP*. Bonn: Galileo Press, 2013.
23. G. Dallas. (2013, May 8, 2018). *Principal Component Analysis 4 Dummies: Eigenvectors, Eigenvalues and Dimension Reduction*. Available: https://georgemdallas.wordpress.

com/2013/10/30/principal-component-analysis-4-dummies-eigenvectors-eigenvalues-and-dimension-reduction/
24. A. Chung. (2018, 8 May 2018). *Supervised Machine Learning—Dimensional Reduction and Principal Component Analysis*. Available: https://hackernoon.com/supervised-machine-learning-dimensional-reduction-and-principal-component-analysis-614dec1f6b4c
25. W. H. Dutton and G. Blank, "Cultures of the Internet: The Internet in Britain - Oxford Internet Survey 2013 Report," 2013, Available: http://oxis.oii.ox.ac.uk/wp-content/uploads/sites/43/2014/11/OxIS-2013.pdf.
26. E. Witteveen *et al.*, "Increased early systemic inflammation in icu-acquired weakness; A prospective observational cohort study," *Critical care medicine,* vol. 45, no. 6, pp. 972–979, 2017.
27. Z. Zhang and A. Castelló, "Principal components analysis in clinical studies," *Annals of translational medicine,* vol. 5, no. 17, 2017.

Chapter 4
Healthcare Analytics Applications

Abstract This chapter provides an overview of many descriptive, predictive, and prescriptive analytics applications in healthcare. Specific algorithms are chosen to illustrate each type of analytic approach. In descriptive analytics, we cover simple descriptive statistics. In predictive analytics, we cover

1. A detailed logistic progression application to illustrate regression analysis
2. Three applications covering decision trees, Naïve Bayes, and natural language processing, to illustrate classification techniques
3. Two applications using *K*-means and hierarchical clustering to illustrate clustering techniques
4. One dimensionality reduction application to illustrate dimension reduction techniques

Finally, one application illustrates the nascent prescriptive analytics. At the end of the chapter, a set of statistical tools is provided.

Keywords Report · Pivot table · Hospital readmission · LACE index · Chart review · Natural language processing (NLP) · *K*-means · Cluster analysis · Rehabilitation service · Principle component analysis (PCA) · Chronic pain

Objectives
At the end of this chapter, you will be able to:

1. Understand what advantages descriptive, predictive, and prescriptive analytics can offer
2. Understand which situations call for which type of analytics
3. Cite one application example for each type of analytics

C. El Morr, H. Ali-Hassan, *Analytics in Healthcare*, SpringerBriefs in Health Care Management and Economics, https://doi.org/10.1007/978-3-030-04506-7_4

4.1 Introduction

Managers make decisions every day; however, for decisions to be evidence-based, they need to rely on data analysis. A decision-making process can be described in four phases: intelligence, design, choice, and implementation [1, 2]. The *intelligence* phase consists of observing reality and acquiring information through data analysis. The *design* phase consists mainly of developing decision criteria and decision alternatives. During the *choice* phase, the decision maker(s) evaluate the different decision alternatives and recommend actions that best meet the decision criteria. During the *implementation* phase, the decision maker(s) might reconsider the decision evaluations as they weigh the consequences of the recommendations; they would then develop an implementation plan, secure the necessary resources and execute the plan. A feedback loop exists between implementation and intelligence, as decisions are implemented, new intelligence/insight may arise that requires a new round of design, choice, and implementation [2] (Fig. 4.1).

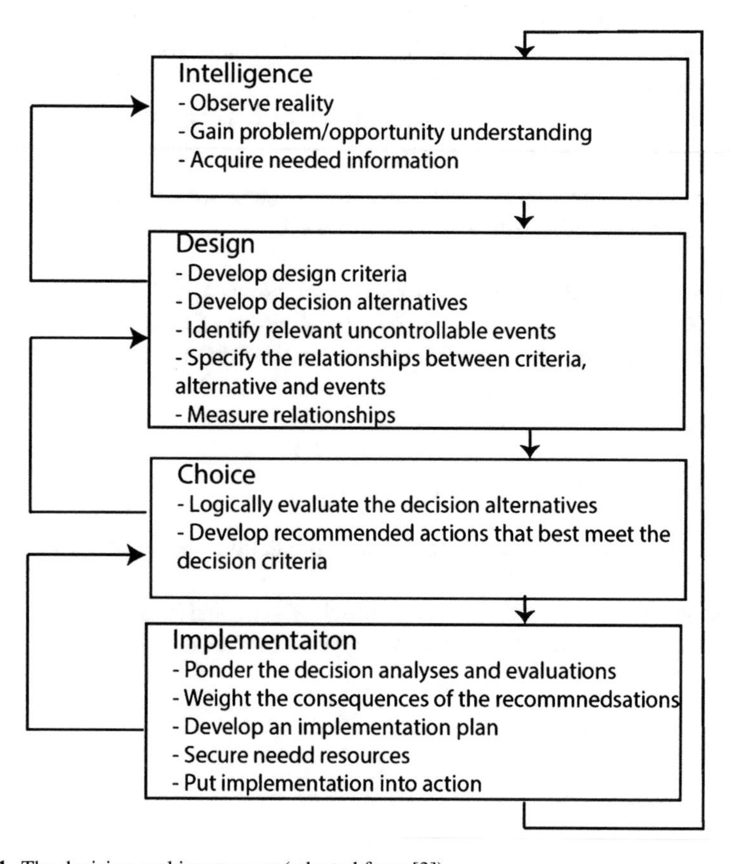

Fig. 4.1 The decision-making process (adapted from [3])

4.2 Descriptive Analytics Applications

Descriptive analytics consider past data analysis to make decisions that help us achieve current and future goals. Statistical analysis is the main "tool" used to perform descriptive analytics; it includes descriptive statistics that provide simple summaries, including graphics analysis, measures of central tendencies (e.g., frequency graphs, average/mean, median, mode), or measures of data variation or dispersion (e.g., standard deviation).

Surveys, interviews, focus groups, web metrics data (e.g., number of hits on a webpage, number of visitors to a page), app metrics data (e.g., number of minutes spent using a feature), and health data stored in electronic records, can be the source of all analytics, including descriptive analytics.

Media companies and social media platforms (e.g., Facebook) use descriptive analytics to measure customer engagement; managers in hospitals can use descriptive analytics to understand the average wait times in the emergency room (ER) or the number of available beds. Descriptive analytics allow us to access information needed to make *actionable decisions* in the workplace. They allow decision makers to *explore* trends in data (why do we have long lines in the ER?), to *understand* the "business" environment (who are the patients coming to the ER?), and to possibly *infer an association* (i.e., a correlation) between an outcome and some other variables (patients with chronic obstructive pulmonary disease tend to have more visits to the ER).

Reports are the main output in descriptive analytics where findings are presented in charts (e.g., bar graph, pie chart), summary tables and the most interesting, pivot tables.

A pivot table is a table that summarizes data originating from another table and provides users with functionality that enables the user to sort, average, sum, and group data in a meaningful way (Figs. 4.2, 4.3, and 4.4).

4.3 Predictive Analytics Applications

In this section of the chapter, we will review how different predictive analytics algorithms are used in healthcare and what they achieve. When appropriate, we will describe four items in each application: the *algorithm* used, the *sample size*, the *timeframe* for the collected data, the *predictors* (input variables), and the *outcome* variable.

4.3.1 Regression Applications

4.3.1.1 Logistic Regression: Predicting 30 Days Hospital Readmission

Hospital readmission is defined as an unplanned admission to a hospital within 30 days of discharge. Hospital readmission has been a major challenge in healthcare worldwide [5, 6]. Since 2012, the Readmission Reduction Program has been used

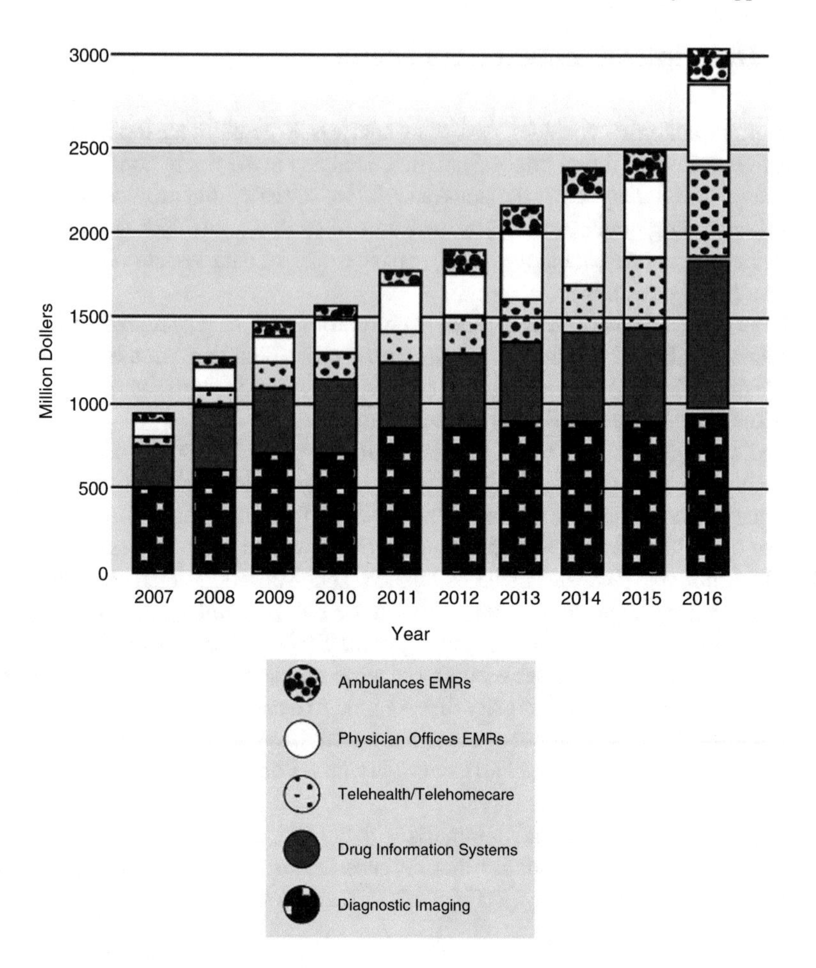

Fig. 4.2 Estimated aggregate benefits in millions of dollars, inflation adjusted to 2016 dollars (Source: Canada Health Infoway [4])

in the United States to measure hospital readmission rates and penalize hospitals with disproportionate rates [7]. In Canada, 8.5% of patients are readmitted within 30 days of discharge [8]. The financial cost resulting from readmission is estimated at $1.8 billion per year [9]. However, it is estimated that 9–59% of unplanned readmissions are preventable [10–12]. Predischarge identification of patients associated with a high risk of readmission could be more cost-effective than postdischarge intervention in reducing readmission rates.

Walraven and colleagues developed the "LACE" index, a cross-conditions tool that predicts early death or unplanned readmission after discharge from the hospital [6]. The authors used a multivariable logistic regression to measure the independent association of these factors with early death or unplanned readmission to hospital and found that four factors impact death or unplanned readmission: length of stay in days (L), acute (emergent) admission (A), comorbidity (Charlson comorbidity

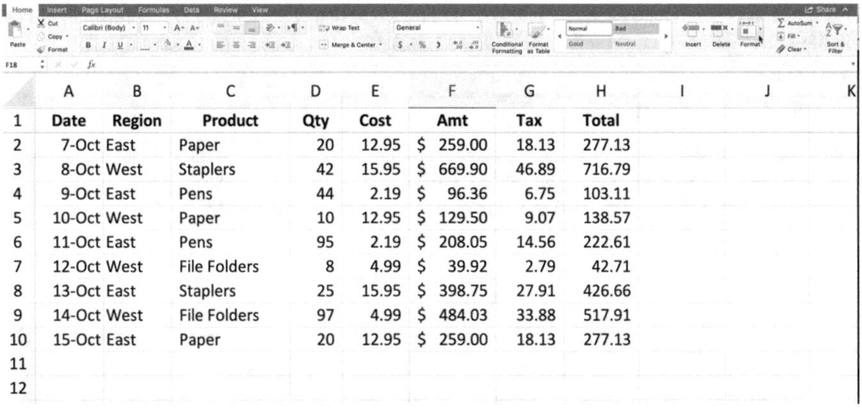

	A	B	C	D	E	F	G	H
1	Date	Region	Product	Qty	Cost	Amt	Tax	Total
2	7-Oct	East	Paper	20	12.95	$ 259.00	18.13	277.13
3	8-Oct	West	Staplers	42	15.95	$ 669.90	46.89	716.79
4	9-Oct	East	Pens	44	2.19	$ 96.36	6.75	103.11
5	10-Oct	West	Paper	10	12.95	$ 129.50	9.07	138.57
6	11-Oct	East	Pens	95	2.19	$ 208.05	14.56	222.61
7	12-Oct	West	File Folders	8	4.99	$ 39.92	2.79	42.71
8	13-Oct	East	Staplers	25	15.95	$ 398.75	27.91	426.66
9	14-Oct	West	File Folders	97	4.99	$ 484.03	33.88	517.91
10	15-Oct	East	Paper	20	12.95	$ 259.00	18.13	277.13

Fig. 4.3 Example of a data sheet

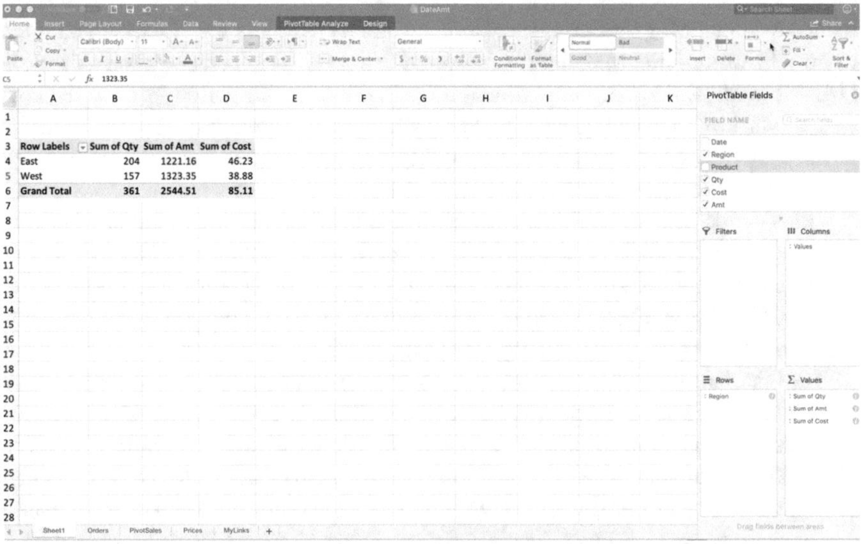

Row Labels	Sum of Qty	Sum of Amt	Sum of Cost
East	204	1221.16	46.23
West	157	1323.35	38.88
Grand Total	361	2544.51	85.11

Fig. 4.4 Example of a pivot table that summarizes data from Fig. 4.3. On the right, we can notice the pivot table fields that allow users to control which summaries are being computed and displayed

index score) (C), and number of visits to emergency department during the previous 6 months (E). The odds ratios associated with the four factors are shown in the following tables (Table 4.1).

An odds ratio of 1.47 for L means that the odds of death or readmission are 1.47 times higher after a one unit (day) increase in the length of stay. The same odds increase 1.84 times for acute admissions, 1.21 times for each one unit increase in the Charlson comorbidity score, and 1.56 times for each additional visit to the emergency department during the previous 6 months.

Table 4.1 Logistic regression result showing the risk of death or unplanned readmission within 30 days after discharge (adapted from Walraven et al. [6])

Variable	Odds ratio (95% CI)
Length of stay in days (logarithm)	1.47 (1.25–1.73)
Acute (emergent) admission	1.84 (1.29–2.63)
Comorbidity (Charlson comorbidity index score)	1.21 (1.10–1.33)
Visits to emergency department during previous 6 months	1.56 (1.27–1.92)

Note: *CI* confidence interval

Had the authors stopped at this point they would have proven the impact of the four factors on the risk of readmission, but they would not have produced a practical tool to be used in a hospital to predict the risk of readmission. However, the authors took one additional step and used a method described by Sullivan and colleagues [13] to transform the logistic regression results into a point system called the LACE index (see next table). The LACE index is comprised of data on the "**L**ength of stay" in the hospital during the current admission, "**A**cuity of admission" (acute/unplanned or not), "**C**omorbidity of patient" (measured using the Charlson comorbidity index) [14, 15], and "**E**mergency department use" in the 6-month period prior to the current admission. In teaching settings, Walraven et al. reported that a 1-point increase in the LACE score increased the probability of an unplanned readmission by 18% and the probability of early death by 29% [6].

Other work, also in teaching settings, found that patients identified as high-risk patients using the LACE tool (LACE score $\geq$ 10) were readmitted twice as often as other patients and had slightly longer lengths of stay [16]. Mixon et al. reported that the LACE index is a better predictor of readmission than measures of patient self-reported preparedness for discharge [17] (Table 4.2).

The Charlson comorbidity score (C) is calculated as follows: 1 point for a history of myocardial infarction, peripheral vascular disease, cerebrovascular disease or diabetes without complications; 2 points for congestive heart failure, chronic obstructive pulmonary disease, mild liver disease or cancer, diabetes with end-organ damage, and any tumor (including lymphoma or leukemia); 3 points for dementia or connective tissue disease; 4 points for moderate to severe liver disease or HIV infection; and 6 points for metastatic cancer.

Other tools addressing hospital readmission such as the UK Nuffield trust model [18] and the Scottish Patients at Risk of Readmission and Admission (SPARRA) [19] exist. The UK Nuffield model was developed in the UK to identify patients with the highest risk of emergency admission and is based on 88 variables extracted from complete hospital and GP systems. SPARRA is a predictive risk stratification tool developed in Scotland to evaluate a person's risk of being admitted to the hospital as an emergency inpatient within the next year. SPARRA holds promise for (a) jurisdictions where resources are devoted to a preventive approach to patient management across the health system and (b) health systems with linked data sets from general practice, home and community care settings, pharmacies, and other settings that allow risk scores to be calculated for large portions of a population [20]. Many

Table 4.2 LACE score calculation [6]

Attribute	Value	Points
Length of stay during the current admission	<1	0
	1	1
	2	2
	3	3
	4–6	4
	7–13	5
	≥14	7
Acute (emergent) admission	Yes	3
	No	0
Comorbidity (Charlson comorbidity index score)	0	0
	1	1
	2	2
	3	3
	≥4	5
Emergency visit (within the last 6 months)	0	0
	1	1
	2	2
	3	3
	≥4	4

jurisdictions continue to face considerable barriers to this level of system and data integration. In such jurisdictions, focusing on reducing readmission using the LACE-rt index remains viable. Both the original LACE index and the LACE-rt index scores range from 0 to 19, where a higher score indicates an increased chance of readmission or early death.

4.3.2 Classification Application

4.3.2.1 Natural Language Processing Application: Automated Chart Review for Asthma

Asthma is one of the most common chronic childhood diseases and is considered to be one of the five most burdensome diseases in the United States. Manual review of patient charts remains the most accurate method to identify asthma cases despite considerable advances in automated approaches, such as natural language processing (NLP), which has been used to mine free-text data in electronic medical records (EMRs); however, an automated approach to detecting the disease is needed for large-scale studies where manual detection through chart review becomes onerous and costly.

The asthma predictive index (API) is a validated criterion to objectively assess asthma status and reduce variability in asthma assessment, such as (1) inconsistent

Table 4.3 Asthma predictive index

Major criteria	Minor criteria
1. Physician diagnosis of asthma for parents	1. Physician diagnosis of allergic rhinitis for patient
2. Physician diagnosis of eczema for patient	2. Wheezing apart from colds
	3. Eosinophilia (≥4%)

asthma criteria (physician diagnosis vs. subjective determination based on diverse asthma criteria) and (2) ascertainment processes (chart review vs. surveys) [21] (Table 4.3).

Kaur et al. explored the potential of NLP methods to classify asthma status at the patient level through mining EMRs, particularly to enable automated chart review for EHR to determine asthma status in children based on API. Researchers designed a cross-sectional study to validate an NLP algorithm for determining asthma status based on API [21].

4.3.2.2 Sample

The study sample consisted of two main cohorts of children. The first training cohort ($n = 87$) was used to train the NLP algorithm (NLP-API); it included children born between 1998 and 2002 just after the implementation of the EHR at the Mayo Clinic.

The second validation cohort ($n = 427$) consisted of a random sample of the 2002–2006 population-based birth cohort who had been enrolled in a previous asthma study and had medical records mainly at the Mayo Clinic. It was used to test and validate the NLP algorithm.

4.3.2.3 Input Attributes

The API index criteria were extracted from many sources: patient-provided information, free-text physician data, and laboratory data. The parents' asthma information (major criteria) was identified from the family history section of the patient's chart and patient-provided information. Eosinophil values were extracted from the lab data when present. Finally, the diagnosis of eczema (major criteria), allergic rhinitis (minor criteria), and wheezing (minor criteria) were extracted from clinical notes.

4.3.2.4 Training Phase

Researchers developed expert rules to classify patients based on the API criteria. The algorithm was developed using the open-source NLP pipeline MedTagger (https://sourceforge.net/projects/ohnlp/files/MedTagger/) developed by the Mayo Clinic and used the training cohort data (Fig. 4.5).

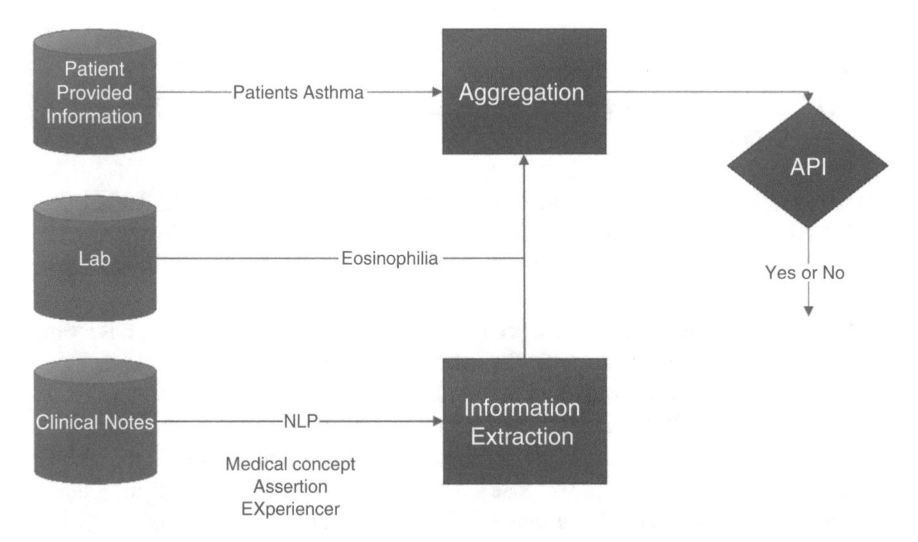

Fig. 4.5 Overview of NLP-API algorithm (adapted from [21])

4.3.2.5 Validation Phase

The asthma status of the validation cohort was determined based on API and using a manual chart review (CW), and the results were compared to their status based on the NLP-API. In other terms, this process measures criterion validity that reflects the extent to which a measurement (e.g., NLP-API classification) is related to an outcome (e.g., diagnosis of asthma).

Researchers calculated the agreement rate, Kappa index, sensitivity, specificity, positive predictive value (PPV), and negative predictive value (NPV) for concordance between NLP-API and manual chart review in terms of asthma status.

4.3.2.6 Results

The final algorithm developed by researchers was able to predict asthmatic status in patients with a sensitivity value of 86% and a specificity value of 98%. Furthermore, the algorithm had a positive predictive value of 88% and a negative predictive value of 98%. Study findings suggest that the NLP algorithm for API mining from EHR was feasible and valid. Future development of the NLP-API algorithm could support population health surveillance and large-scale clinical studies focusing on asthma [21].

4.3.3 *Clustering Application*

4.3.3.1 *K*-Means Application: Cluster Analysis of Rehabilitation Service Users [22]

Canada's aging demographics have resulted in increased pressure to limit additional costs to the healthcare system. One of the primary initiatives funded by the Ontario government in recent years to address this burden is increased accessibility to home-based care.

Rehabilitation therapies have demonstrated effectiveness in enabling autonomy and independence for seniors receiving care in the home setting.

Research in the field focused on measuring health and service needs by identifying predictors and risk factors leading to service use, institutionalization, functional improvement/decline, and mortality of the aging population. However, the elderly have differing multiple chronic diseases, medical conditions, and prescription medications. Hence, researchers have suggested exploring patient heterogeneity to obtain unbiased model outcomes when making uniform policy recommendations for a heterogeneous aging population by using the *K*-means algorithm to uncover previously unidentified patterns of clinical characteristics and to create client profiles for different subgroups of elderly receiving homecare.

Researchers sought to explore the heterogeneity of homecare recipients who receive rehabilitation therapies using a *K*-means cluster analysis. They used data collected between April 2005 and August 2008 from the Resident Assessment Instrument—Home Care (RAI-HC) and a variety of assessment items, including demographic information, service use, and physical function. The data of 150,253 persons who had received rehabilitation services within the first 3 months of their assessment were used in this analysis. The study also used service outcome data (e.g., hospitalization, mortality, successful completion of homecare plans) that were collected a year after each client's first homecare assessment [22].

All variables, except for age, were dichotomized (transformed into two categories) for the cluster analysis. Researchers used the FASTCLUS procedure in SAS version 9.1 to conduct the analysis. The input of the algorithm consisted of 37 variables and included the following:

1. Activities of daily living (ADLs)
2. Instrumental activities of daily living (IADLs)
3. Disease diagnoses
4. Changes in health
5. Health instability outcome measures
6. Age

Seven main clusters were identified, and these outcomes are identified below:

1. Dependent and immobile clients with cognitive problems (9%)
2. Dependent but mobile clients with cognitive problems (23%)
3. Primarily women clients requiring assistance with IADLs and some ADLs (14%)

4. Primarily women clients requiring assistance with IADLs (15%)
5. Clients requiring assistance with IADLs and bathing (19%)
6. ADLs independent cognitively intact younger clients (10%)
7. Clients who live alone and require some assistance with housework and bathing (10%)

Each of these clusters has different activities of daily living (ADLs); Changes in Health, End-Stage Disease, and Signs and Symptoms Scale (CHESS) values; instrumental activities of daily living (IADLs); occupational therapy (OT); and physical therapy (PT) needs.

Study findings demonstrated the diversity of the senior population who receive rehabilitation services in the home and how these users can be segmented into homogenous clusters that share key patient traits [22].

4.3.4 Dimensionality Reduction Application

4.3.4.1 Principle Component Analysis (PCA) Application: Delineation Between Different Components of Chronic Pain

In healthcare, pain experiences are traditionally measured using psychometric measures, such as scales and questionnaires. These psychometric measures have considerable overlap because pain-related processes have closely interlinked affective and modulatory processes in the nervous system. To help clarify what relationships exist between neural processes and questionnaire responses, the medical community has often turned to neuroimaging studies to understand the complexities of pain perception. The present study explored the ability of dimension reduction methods, specifically principal component analysis (PCA), to delineate different components of chronic pain. The sample consisted of validated questionnaires and fMRIs (functional magnetic resonance images) of 38 postmenopausal patients who were suffering from osteoarthritis of carpometacarpal joints. Study data originated from previously conducted studies, and no additional consent was required for the current study. Questionnaires to measure psychometrics, personality traits, and pain-related assessment were used including the McGill Pain Questionnaire (pain), the Beck Depression Inventory (depression), the Spielberger State-Trait Anxiety Inventory (anxiety), the Eysenck Personality Questionnaire (personality traits), the Patient-Rated Wrist and Hand Evaluation: PRWHE (hand and wrist pain), and patient MRI data [23].

Dimension reduction was performed using principal component analysis (PCA) and oblique rotation. PCA is a mathematical algorithm that transforms a number of correlated variables into a smaller number of uncorrelated variables called principal components [23].

Two principal components were identified, and they accounted for 73% of the total variance; hence, they explained much of the pain experience. Component 1 accounted for 49.7% of the total variance in the data and was highly related to the

PRWHE scores and to all elements of the McGill pain scores, which denoted similar sources of variance for the PRWHE and McGill scores.

Component 2 was found to be related to psychological traits and accounted for 23.2% of the total variance. Because these are the two components that explain almost 95.2% of the variation in the data, researchers decided to investigate the relationship between these components and the blood flow in the brain instead of investigating relationships between all data that were originally available [23].

4.4 Prescriptive Analytics Application

Prescriptive analytics go beyond prediction to prescribe an optimal course of action to reach a certain goal based on predictions of future events. A simple example would be an app that predicts the time duration of a journey from a current location to certain destinations if the App is equipped with prescriptive analytics, then it can prescribe the shortest path/way to reach the destination after comparing several alternative routes.

4.4.1 Prescriptive Analytics Application: Optimal In-Brace Corrections for Braced Adolescent Idiopathic Scoliosis (AIS) Patients

Brace treatment is a common nonsurgical treatment for adolescent idiopathic scoliosis (AIS). The objective of a brace designer is to achieve the maximum possible in-brace curve correction; however, the target corrections are approximately 50%. Each patient is unique; hence, customizing brace treatment per patient is supposed to enhance the treatment results. A study by Chalmers et al. applied prescriptive analytics to obtain customized recommendations for correction and tested their efficacy.

Retrospective data for 90 AIS patients who had brace treatment (60 full-time and 30 nighttime braces) from 2006 to 2013 were collected. Researchers used a predictive modeling technique that considered the (1) patient age, (2) the Cobb angle (a measure of the curvature of the spine), (3) the scoliometer measurement, and (4) in-brace correction to predict whether the patient's major Cobb angle will progress (become worse), improve, or remain neutral during treatment.

The model predicted progression of the major Cobb angle (progress, improve, or neutral) for in-brace corrections ranging between 20% and 160% for each patient. The known progression rates stored in the patient charts were then compared to the predicted progression rates based on the prescribed (recommended) in-brace corrections.

In 37% of cases, the correction automatically prescribed by the model was less aggressive (i.e., more comfortable and desirable) than the actual correction applied to the patient, without compromising the treatment outcome.

The statistical analysis showed significantly fewer (26% less) progressive cases if one would have followed the prescriptive analytics suggestions in comparison with the actual corrections observed in the charts. One of the most important findings was that the patient whose prescribed correction was less than the applied correction showed an increased progression if the patient had followed the prescriptive model. This study suggests that prescriptive analytics can improve health outcomes for braced adolescent idiopathic scoliosis (AIS) patients.

4.5 Conclusion

Applications of analytics in healthcare are increasing. The wide range of these applications is remarkable, and their accuracy is promising in most cases and astonishing in some cases. The future will bring progress in terms of accuracy and the number of fields where analytics are applied. The advancement of machine learning tools and prescriptive analytics raises many questions regarding the future of some jobs (as we will see in Chap. 6) and the impact on the work processes and health outcomes. The impact has yet to be studied.

References

1. H. A. Simon, *Administrative Behavior, 4th Edition*. Free Press, 1997.
2. L. C. Jain and C. P. Lim, "Advances in Intelligent Decision Making," in *Handbook on Decision Making: Vol 1: Techniques and Applications*, vol. 1, C. P. Lim, Ed.: Springer Berlin Heidelberg, 2010.
3. G. A. Forgionne, "Decision-Making Support System Effectiveness: the Process to Outcome Link %J Inf. Knowl. Syst. Manag," vol. 2, no. 2, pp. 169-188, 2000.
4. Canada Health Infoway, "Year In Review 2016-2017," Canada Health Infoway, Toronto Jul 28, 2017 2017, Available: https://www.infoway-inforoute.ca/en/component/edocman/resources/i-infoway-i-corporate/annual-reports/3350-annual-report-2016-2017.
5. G. F. Anderson and E. P. Steinberg, "Hospital readmissions in the Medicare population," (in eng), *N Engl J Med,* vol. 311, no. 21, pp. 1349–53, Nov 22 1984.
6. C. van Walraven *et al.*, "Derivation and validation of an index to predict early death or unplanned readmission after discharge from hospital to the community," (in eng), *Cmaj,* vol. 182, no. 6, pp. 551–7, Apr 6 2010.
7. Centers for Medicare & Medicaid Services. (2014, July 22). *Readmission Reduction Program*. Available: https://www.cms.gov/Medicare/Medicare-Fee-for-Service-Payment/AcuteInpatientPPS/Readmissions-Reduction-Program.html Accessed: 2017-01-31. Archived at http://www.webcitation.org/6nw9fZ955
8. (2013). *Health Indicators 2013: Definitions, Data Sources and Rationale*. Available: https://www.cihi.ca/en/ind_defin_2013_en.pdf
9. Canadian Institute for Health Information, "All-Cause Readmission to Acute Care and Return to the Emergency Department," CIHI, Ottawa, ON2012, Available: https://secure.cihi.ca/free_products/Readmission_to_acutecare_en.pdf.
10. S. E. Frankl, J. L. Breeling, and L. Goldman, "Preventability of emergent hospital readmission," (in eng), *Am J Med,* vol. 90, no. 6, pp. 667–74, Jun 1991.

11. C. H. Yam *et al.*, "Avoidable readmission in Hong Kong--system, clinician, patient or social factor?," (in eng), *BMC Health Serv Res,* vol. 10, p. 311, Nov 17 2010.
12. The Canadian Medical Protective Association. (2014). *Reducing unplanned hospital readmissions.* Available: https://www.cmpa-acpm.ca/en/duties-and-responsibilities/-/asset_publisher/bFaUiyQG069N/content/reducing-unplanned-hospital-readmissions. Accessed: 2017-01-31. Archived at http://www.webcitation.org/6nw8r6sQz
13. L. M. Sullivan, J. M. Massaro, and R. B. D'Agostino, Sr., "Presentation of multivariate data for clinical use: The Framingham Study risk score functions," (in eng), *Stat Med,* vol. 23, no. 10, pp. 1631–60, May 30 2004.
14. Health Systems Performance Research Network (HSPRN). (2016, Aug. 17). *Online LACE index Tool.* Available: http://www.hospitalreport.ca/?p=33. Accessed: 2017-01-31. Archived at http://www.webcitation.org/6nw9Gdg9y
15. QIO. (2016, Aug. 17). *Online LACE index Tool.* Available: http://qio.ipro.org/care-transitions/healthcare-professionals/past-events/united-healthnd-ac-servicesbinghamton-lace-tool-for-assessment-of-risk-for-read. Accessed: 2017-02-11. Archived at http://www.webcitation.org/6oCSHhyRK
16. A. Gruneir *et al.*, "Unplanned readmissions after hospital discharge among patients identified as being at high risk for readmission using a validated predictive algorithm," *Open Medicine,* vol. 5, no. 2, pp. e104-e111, 2011.
17. A. S. Mixon *et al.*, "Preparedness for hospital discharge and prediction of readmission," (in Eng), *J Hosp Med,* Feb 29 2016.
18. J. Billings, T. Georghiou, I. Blunt, and M. Bardsley, "Choosing a model to predict hospital admission: an observational study of new variants of predictive models for case finding," *BMJ Open,* vol. 3, no. 8, p. e003352, Aug 26 2013.
19. IS Scotland. Scottish patients at risk of readmission (SPARRA). (2017). *SPARRA Model.* Available: http://www.isdscotland.org/Health-Topics/Health-and-Social-Community-Care/SPARRA/SPARRA-Model/ Accessed: 2017-01-30. Archived at http://www.webcitation.org/6nuA2VRlf
20. A. Mahmoud, "Scottish patients at risk of readmission and admission (Sparra)," *International Journal of Integrated Care,* vol. 16, no. 6, p. A216, 2016.
21. H. Kaur *et al.*, "Automated chart review utilizing natural language processing algorithm for asthma predictive index," (in eng), *BMC Pulm Med,* vol. 18, no. 1, p. 34, Feb 13 2018.
22. J. J. Armstrong, M. Zhu, J. P. Hirdes, and P. Stolee, "K-means cluster analysis of rehabilitation service users in the Home Health Care System of Ontario: examining the heterogeneity of a complex geriatric population," (in eng), *Arch Phys Med Rehabil,* vol. 93, no. 12, pp. 2198–205, Dec 2012.
23. D. Keszthelyi *et al.*, "Delineation between different components of chronic pain using dimension reduction - an ASL fMRI study in hand osteoarthritis," (in eng), *Eur J Pain,* Mar 9 2018.

Chapter 5
Data Visualization

Abstract This chapter introduces the concept of data visualization and its role in helping readers understand and interpret data and spot valuable information such as trends and exceptions. The different graphical elements and charts are introduced with examples of healthcare visualizations created using different software tools. This chapter overviews the key relationships in data that can be displayed and highlighted in graphs. The chapter then covers infographics and dashboards, which are visualization-rich tools that are increasingly being used in the healthcare industry. The chapter ends with guidelines for building good visualizations, an outline of software tools that can be used for creating visualizations, a conclusion, and a list of references.

Keywords Data visualization · Graphs · Charts · Graphical objects · Relationships · Infographics · Dashboards

Objectives

At the end of this chapter, you will be able to:

1. Understand how data are visualized via graphs and charts
2. Identify the basic objects in a graph and their characteristics
3. Understand how different charts can be used to represent different types of relationships embedded in the data
4. Understand how infographics can be used for storytelling and dashboards for monitoring and managing an organization
5. Identify some key guidelines for effective visualizations
6. Understand how analytics software supports data visualizations

C. El Morr, H. Ali-Hassan, *Analytics in Healthcare*, SpringerBriefs in Health Care Management and Economics, https://doi.org/10.1007/978-3-030-04506-7_5

5.1 Introduction

Visualization via charts, graphs, and images is an effective and efficient way to interpret and understand data and help spot valuable information such as patterns, trends, and anomalies [1]. The reason is that graphs, unlike tables and written text, are primarily visual in nature, and approximately 70% of our sense receptors are dedicated to vision [2].

While the invention of data visualization may not be easily attributed to one individual, William Playfair (1759–1823) is generally viewed as the inventor of many common graphical forms, such as bar and pie charts [3]. Of his well-known visualizations is his balance of trade and chart of the national debt of England (Fig. 5.1).

Another classic example of an old visualization is the illustration of Napoleon's failed Russian campaign of 1812 by Charles Minard (Fig. 5.2). The graph displays the number of French soldiers marching towards and then retreating from Moscow, overlaid on top of a map. The thickness of the band is representative of the number of soldiers, which decreased as the army moved from France on the right to Russia on the left. Underneath the map is a line chart displaying the temperature that soldiers faced as they moved during the campaign.

In the remainder of this chapter, we introduce the basics of data visualizations, including a taxonomy of basic graphical objects and charts and their uses. We include a number of healthcare visualizations we generated with three different software

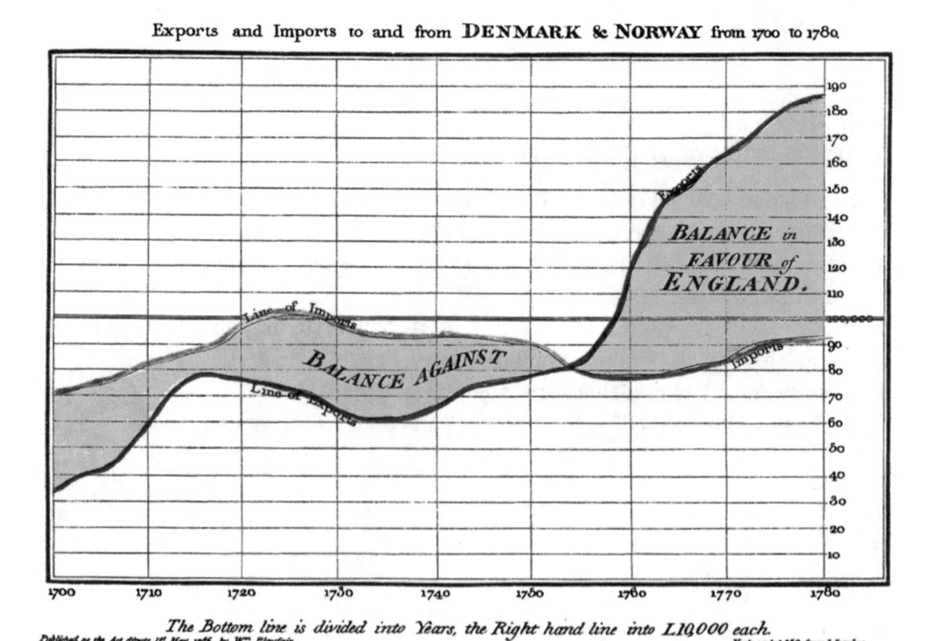

Fig. 5.1 William Playfair's balance of trade and chart of national debt of England

packages using different sources of open data. We also cover infographics and dashboards, which are visualization-rich tools that are increasingly being used in the healthcare industry. We finish with guidelines for building good visualizations.

5.2 Presentation and Visualization of Information

The type of data sometimes dictates the type of graph that can or cannot be used. As a quick reminder, data are broken into two types: quantitative and categorical. Quantitative values measure things and consist of a quantity and unit of measure (e.g., 300 km). Categorical data divide information into useful groups and are nominal (e.g., fall, winter, spring, summer), ordinal (e.g., low, medium, high), interval (e.g., 0–9, 10–19, …), and hierarchical (e.g., year, quarter, month, week, day). In many cases, the type of data dictates the type of graph and visualization to be used.

5.2.1 *A Taxonomy of Graphs*

Of the different available taxonomies of graphs, we will follow the one proposed by Stephen Few [2], a well-known expert in data visualization. According to Few, quantitative data can be basically represented in graphs by the following six basic objects: points, lines, bars, boxes, shapes with varying 2-D areas, and shapes with varying color intensity.

A *point* is a simple dot on a graph representing two values, one on each axis and is referred to as a scatterplot (Fig. 5.3). Scatterplots are used when the values do not start at zero. They give a general idea about the distribution of the data and are useful in highlighting relationships such as correlations and in detecting data outliers [1, 2].

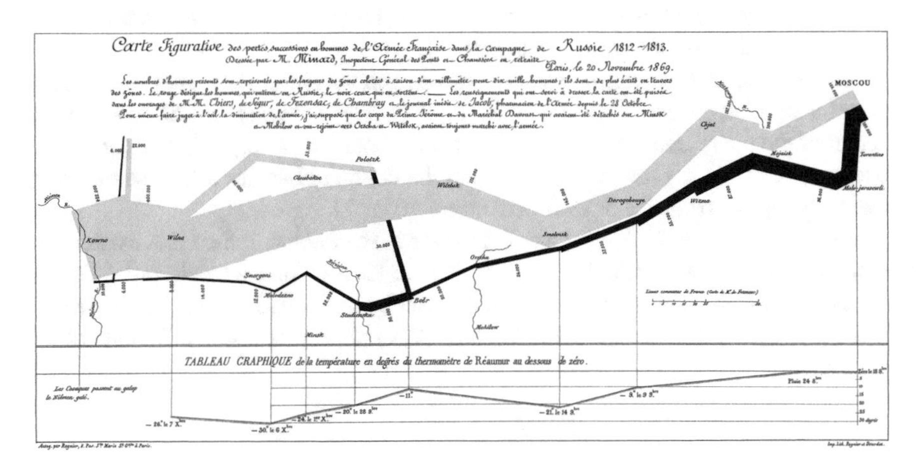

Fig. 5.2 Napoleon's failed Russian campaign of 1812 by Charles Minard

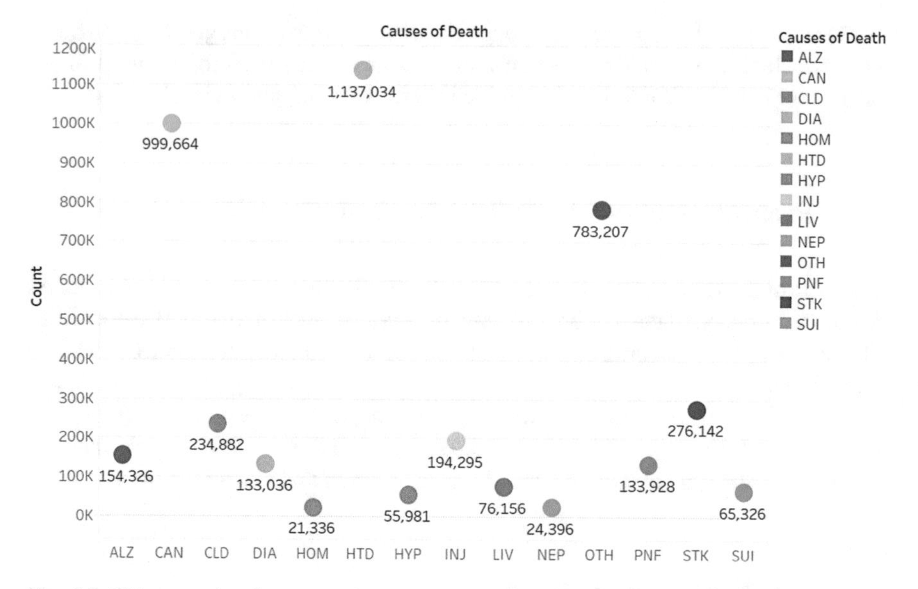

Fig. 5.3 This example of a *scatterplot* shows the total number of deaths by different causes in California from 2012 to 2016. With a scatterplot, you can easily and quickly identify, for example, the top three causes of deaths and their numbers. This graph was generated using Tableau software and the California Department of Public Health Open Data [4]

A *line* connects a series of values in a graph and is a good representation of how values change over time (called a time series) [1] (Fig. 5.4). It is also used to display a trend in a scatterplot (Fig. 5.5) [2].

A *bar* is a rectangle that encodes quantitative information by its length. Bars are easy to see and compare and should always begin at the value zero [2]. Vertical bars are referred to as columns (Fig. 5.6).

A *box* is also rectangular but encodes a wide range of values, such as the minimum, maximum, and median values (Fig. 5.7). They are used most often to compare the distribution of different data sets [2].

Shapes with 2-D areas represent values in proportion to their area rather than their location on the graph. A popular example is the *pie chart* (Fig. 5.8), where each sector of the pie represents a percentage of the whole. However, despite its frequent use, a pie chart is not recommended when the compared values are close or when there are many categories or sectors to compare [1, 2].

Another example of shapes with 2-D areas is the *bubble,* which is a scatterplot that quantifies three values, two by their relative location on each axis and the third by the size of the bubble. A fourth variable can be quantified by applying *variable intensities of the same color* to the bubbles [2] or simply by using different colors (Fig. 5.9).

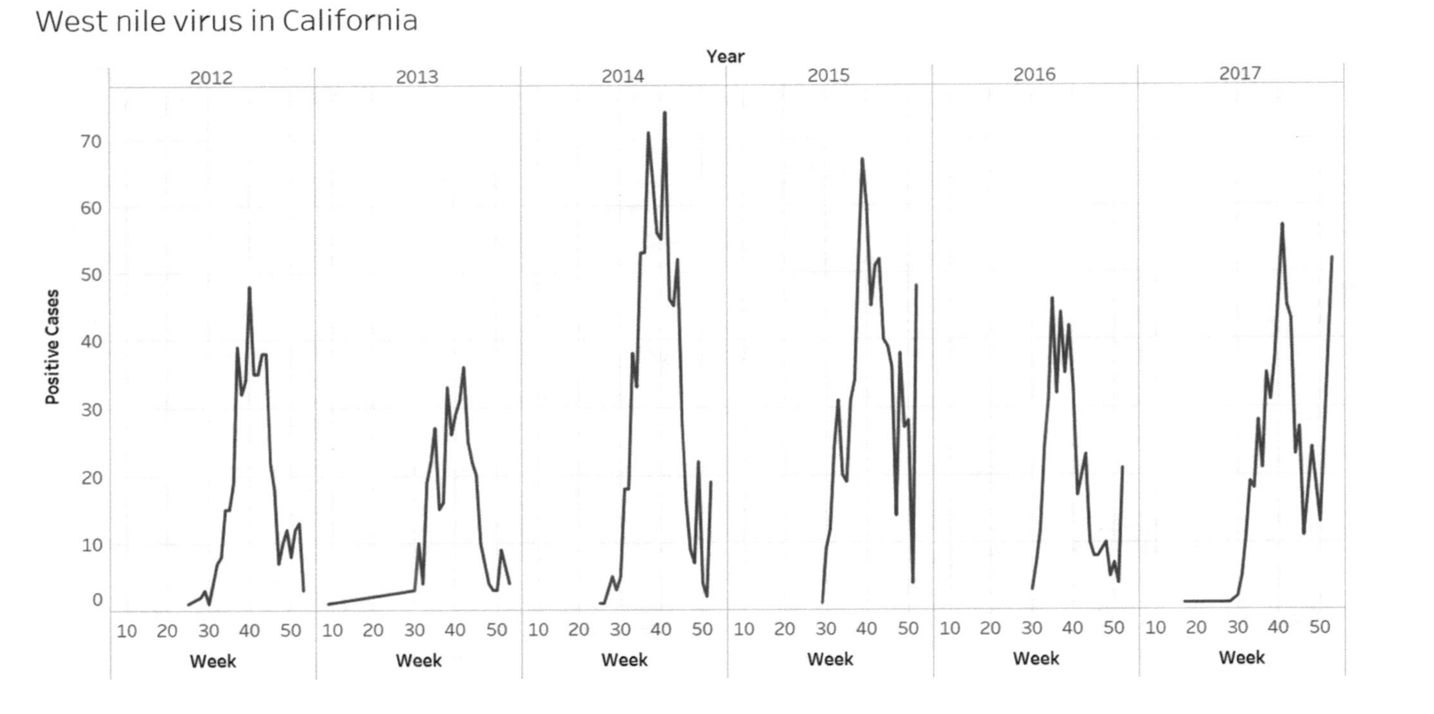

Fig. 5.4 This example of a *line chart* displays the incidence of the West Nile virus in California on a weekly basis from 2012 to 2017. It clearly shows a yearly cycle where the incidence peaks between weeks 39 and 42. This graph was generated using Tableau software and the California Department of Public Health Open Data [5]

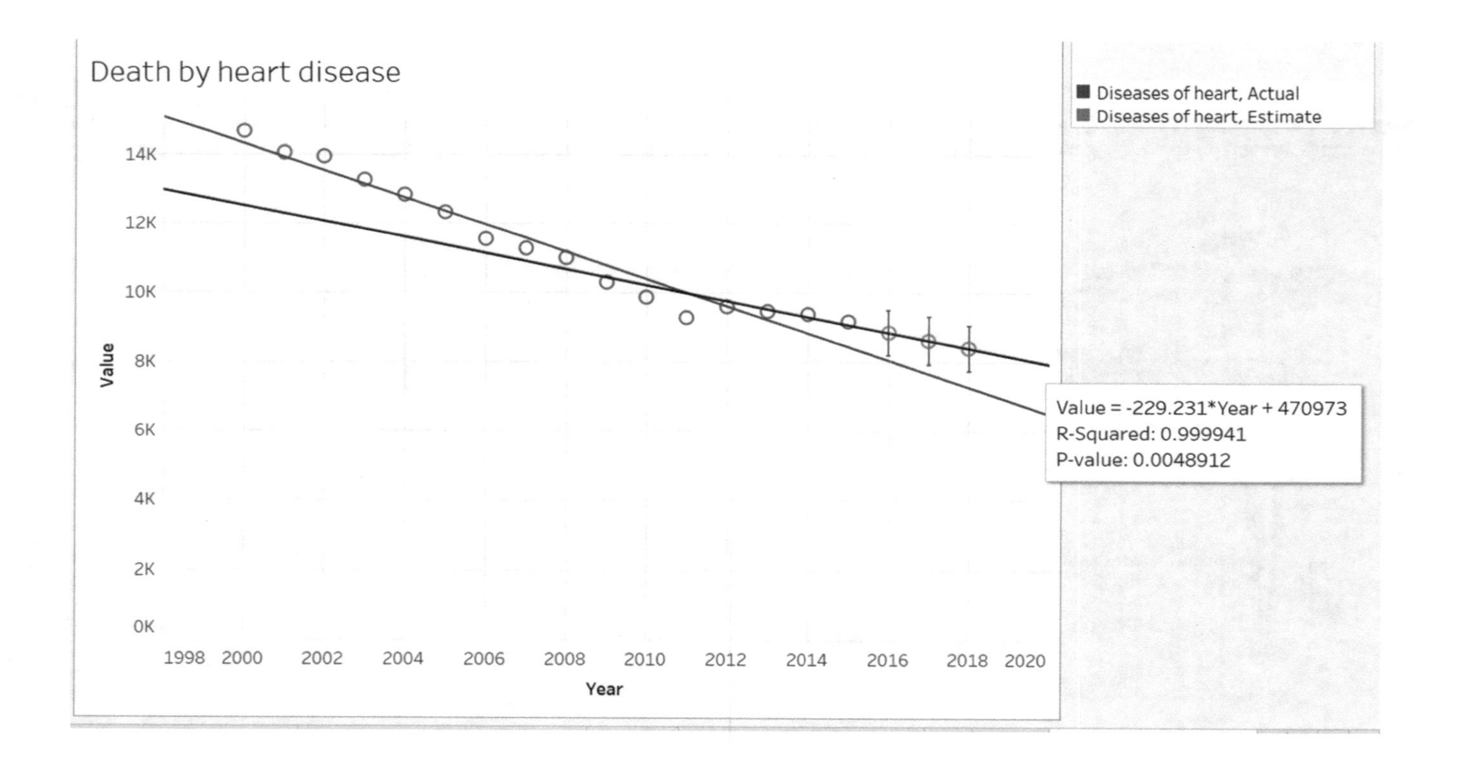

Fig. 5.5 This example of a *scatterplot with a line* highlighting the clear trend of decreasing death by heart disease in Canada from 2000 to 2016. The software generated a regression line statistic along with a projection until 2020. This graph was generated using Tableau software and Statistics Canada data [6]

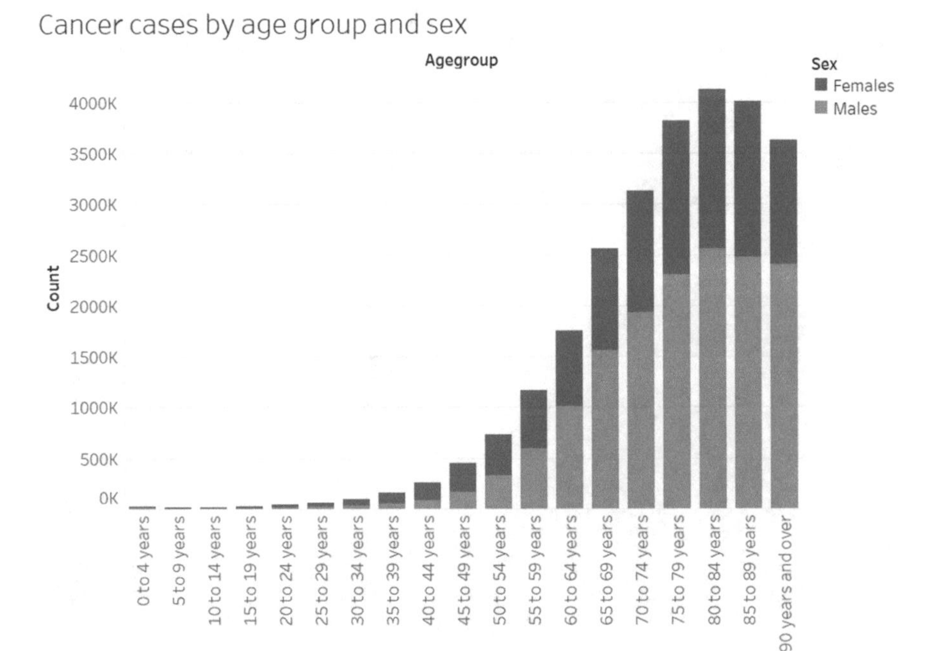

Fig. 5.6 This example of a *bar graph*, also called a column chart, displays the number of cancer cases in Canada by age group and sex (1992–2015). It is called a stacked bar chart because it is a combination of more than one data set [1]. The graph clearly highlights which age groups and sex are more susceptible to the disease and shows a pattern or relationship between cancer, age, and sex. This graph was generated using Tableau software and Statistics Canada data [7]

5.2.2 *Relationships and Graphs*

Graphs are used to display relationships in data by giving them shapes. There are eight main types of relationship graphs that are typically used: time series, ranking, part-to-whole, deviation, distribution, correlation, geospatial, and nominal comparison [2]. *Time series* graphs show how something changed (increased, fluctuated, declined…) over time (e.g., Figs. 5.4 and 5.5). Graphs display *ranking* relationships such as larger than, smaller than, and equal to, sorted in increasing or decreasing order (Fig. 5.6 though not sorted). Graphs display *part-to-whole* relationships by showing how individual values make up the whole of something (for example, by percentage or rate of total) and how they compare to each other (Figs. 5.6 and 5.8). *Deviations* represent how one or more sets of values differ from a reference set of values (Fig. 5.10) [2].

A *distribution* represents how values are distributed across an entire range, from the lowest to the highest and is called a *frequency distribution* when it shows the number of times something occurs. When bars are used, it is referred to as a histogram (Fig. 5.11) [2].

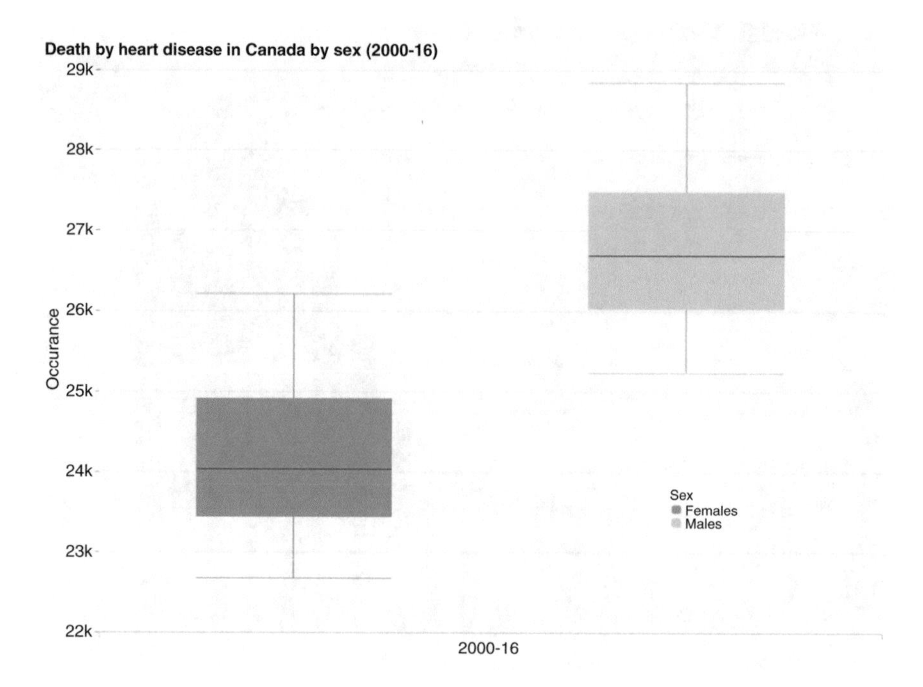

Fig. 5.7 This example of a box graph, known as a *box plot*, represents death by heart disease by sex in Canada (2000–2016). Each box displays the minimum, first quartile, median, third quartile, and maximum values in the data set. This graph was generated using SAP Lumira software and Statistics Canada data [6]

A graph displays a *correlation* when it shows whether two sets of values vary (increase, decrease, follow) in relation to each other, positively or negatively and to what degree (e.g., Figs. 5.5, 5.6, and 5.10). G*eospatial* relationships between values are displayed by plotting them on a map (Fig. 5.12). Finally, a *nominal comparison* is the simple display of a set of discrete quantitative values so that they can be easily read and compared (e.g., Fig. 5.3) [2].

To display a specific relationship graphically, different objects and types of graphs can be used, with some being more adequate for the task than others while others should be avoided. Table 5.1 is a summary of the recommended graphical objects used to display each type of relationship described above.

In addition to the visualizations shown so far, there is an infinite number of advanced visualizations that are very popular and useful. Below are some additional examples of popular or interesting advanced visualizations.

There are infinite additional ways to visualize data and information. What has been covered so far in this chapter is an introduction to the basic and most popular visualizations. To view additional examples of interesting and rich visualizations, you can explore numerous sources such as the Information is Beautiful website (https://informationisbeautiful.net/) and Tableau's public gallery (https://public.tableau.com/en-us/s/gallery). A number of excellent interactive healthcare visualizations can be found at the website of The Institute for Health Metrics and Evaluation (IHME),

Table 5.1 Graphical object to use for each type of relationship (adapted from Few [2])

	Graphical objects			
Relationship	Points	Lines	Bars	Boxes
Time series (categorical data on the x-axis and quantitative values on the y-axis)	Dot plot only when values were not collected at consistent intervals of time	For emphasis on the overall pattern (e.g., Figs. 5.4 and 5.5)	For emphasis on individual values	Only when showing distributions that change through time
Ranking	Dot plot when the quantitative scale does not start at zero; otherwise, use bar	Avoid	Horizontal or vertical (e.g., Fig. 5.6 but preferably sorted)	Only when ranking multiple distributions. Horizontal or vertical
Part-to-whole	Avoid	To display how parts of a whole change over time	Horizontal or vertical (e.g., Fig. 5.6)	Avoid
Deviation	Dot plot when the quantitative scale does not start at zero (e.g., Fig. 5.10)	Useful when combined with time series	Horizontal or vertical. Always vertical if combined with time series	Avoid
Distribution (single)	Known as a strip plot. Emphasis on individual values	Known as a frequency polygon. Emphasis on overall pattern	Known as a histogram. Emphasis on individual intervals (e.g., Fig. 5.11)	Avoid
Distribution (multiple)		Known as a frequency polygon. Limit to a few lines	Avoid	Known as a box plot (e.g., Fig. 5.7)
Correlation	Known as a scatterplot (e.g., Fig. 5.5)	Avoid	Horizontal or vertical	Avoid
Geospatial	Different point sizes encode values (e.g., Fig. 5.12)	Used to mark routes (e.g., Fig. 5.13)	Avoid	Avoid
Nominal comparison	Use a dot plot if scales do not start at zero (e.g., Fig. 5.3)	Avoid	Horizontal or vertical. Use when the scale starts at zero (e.g., Fig. 5.6)	Avoid

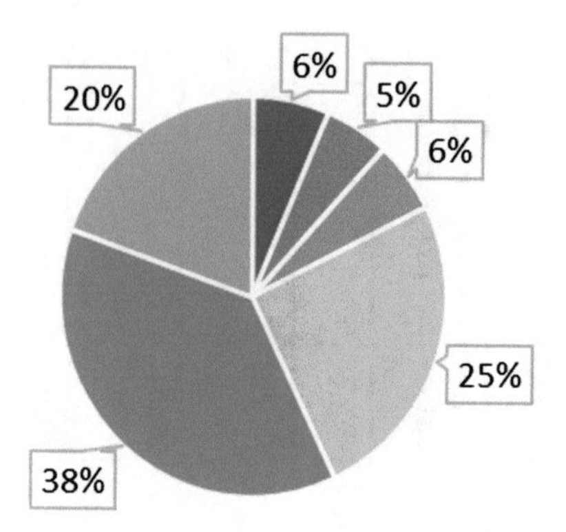

Fig. 5.8 This example of a *pie chart* depicts the proportion of the top six causes of death of Canadian males (2000–2016). Note that the percentages add up to 100% of the deaths by the six causes and exclude the many other causes of death. This graph was generated using Microsoft Excel software and Statistics Canada data [6]

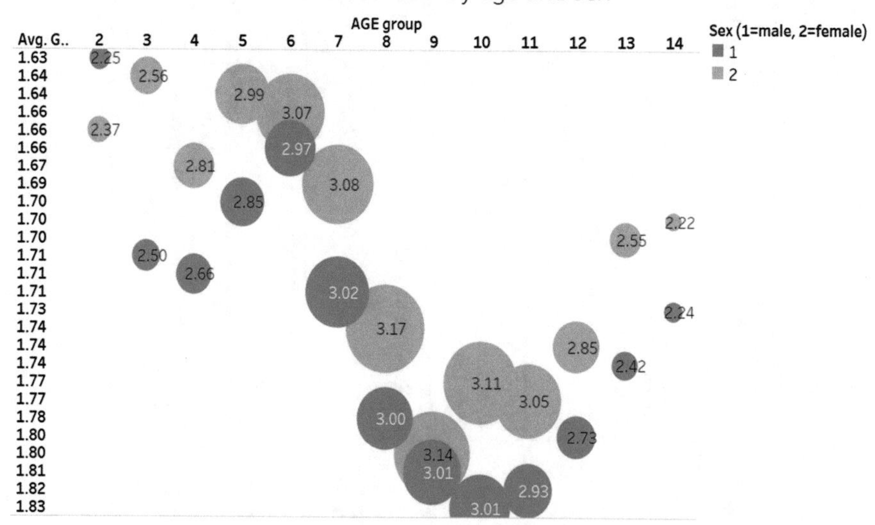

Fig. 5.9 This example of a *bubble chart* displays the average life satisfaction level (*Y*-axis) and work stress level (bubble size), by age group (*X*-axis) and sex (color) in Canada (2012). Among the visual observations are that after a certain age, females have a lower life satisfaction and higher stress level but that there is no clear relationship between life satisfaction and work stress in general. This graph was generated using Tableau software and Statistics Canada data [8]

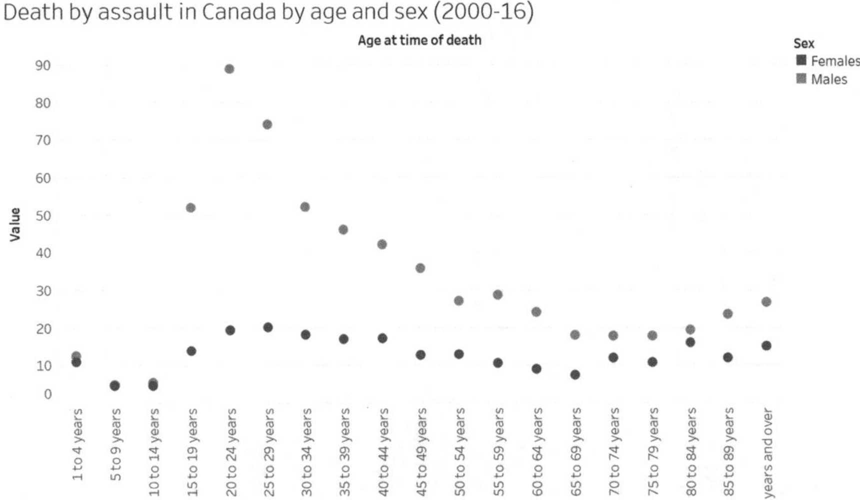

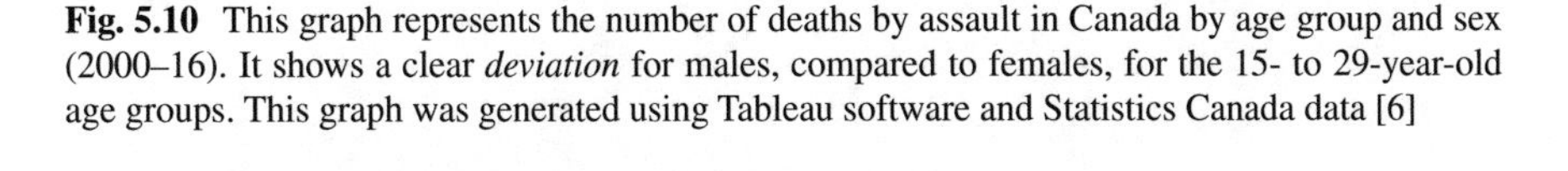

Fig. 5.10 This graph represents the number of deaths by assault in Canada by age group and sex (2000–16). It shows a clear *deviation* for males, compared to females, for the 15- to 29-year-old age groups. This graph was generated using Tableau software and Statistics Canada data [6]

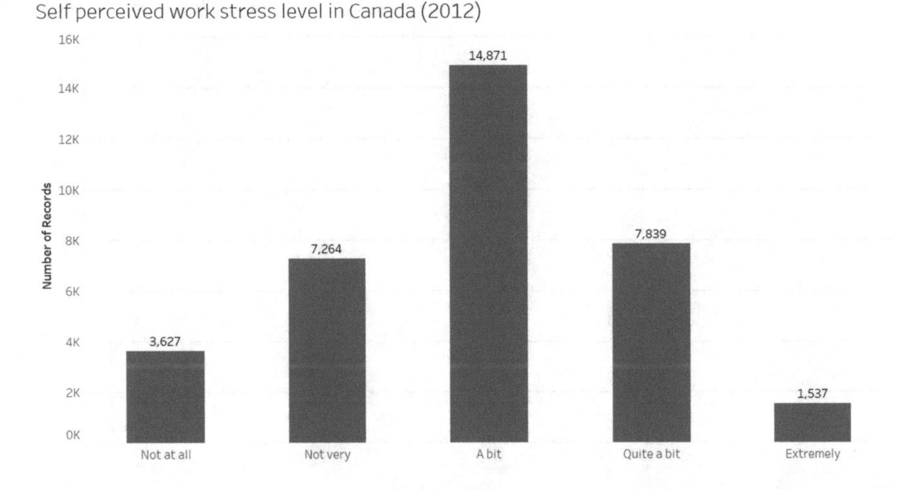

Fig. 5.11 This example of a *histogram* displays the *distribution* of the five levels of self-perceived work stress in Canada in a 2012 survey. The highest number of respondents reported that they feel "a bit" stressed at work, while the lowest number felt "extremely" stressed. This graph was generated using Tableau software and Statistics Canada data [8]

which is an independent global health research center at the University of Washington (http://www.healthdata.org/results/data-visualizations). Two examples can be found in Figs. 5.13 and 5.14. Most visualizations are interactive with filters that allow the viewer to select from a variety of graph types, regions, dates, diseases, etc. Some are also dynamic and display data evolving over a period of time.

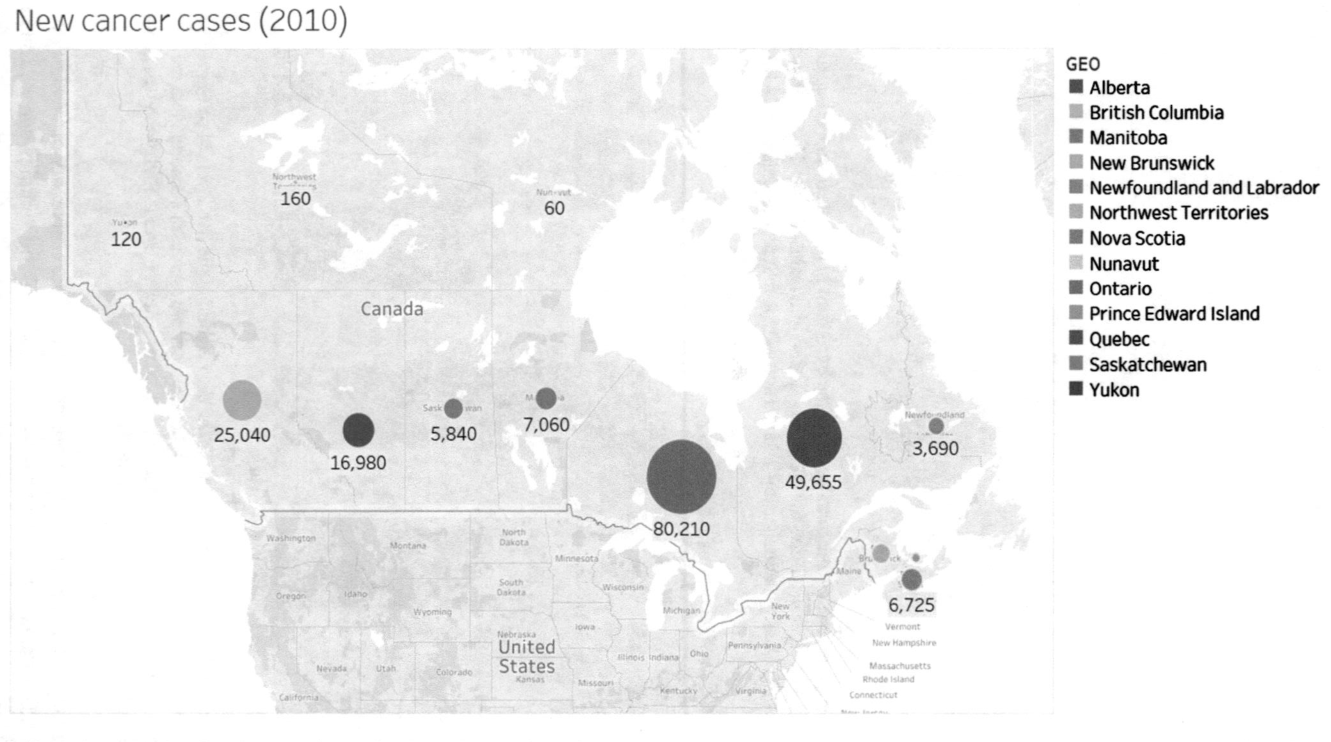

Fig. 5.12 This example of a *geospatial map* displays the number of new cancer cases in the different Canadian provinces and territories in 2010. This graph was generated using Tableau software and Statistics Canada data [7]

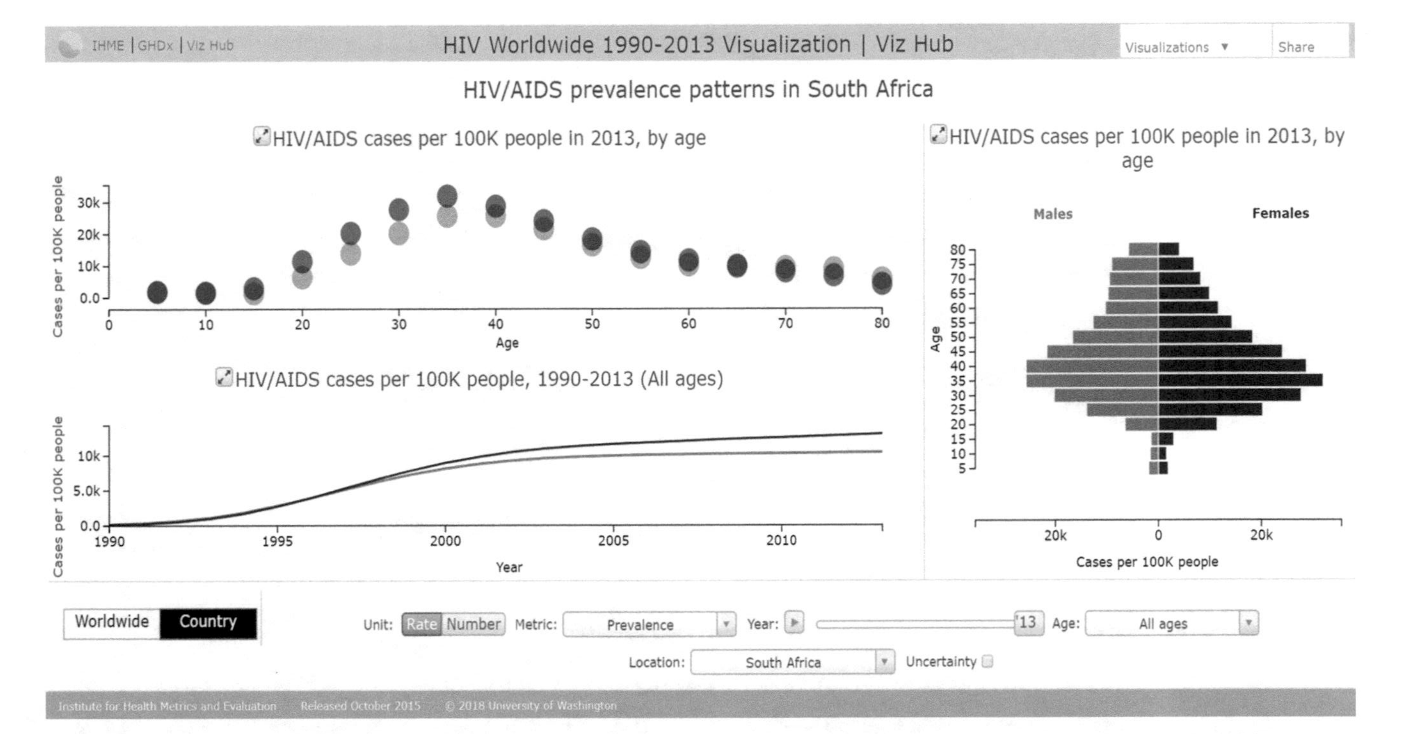

Fig. 5.13 This is an example of a scatterplot, time series, and pyramid graphs. It displays patterns of HIV/AIDS prevalence in South Africa by sex and age from 1990 to 2013. The graph was generated on the Health Metrics and Evaluation (IHME) website [9]

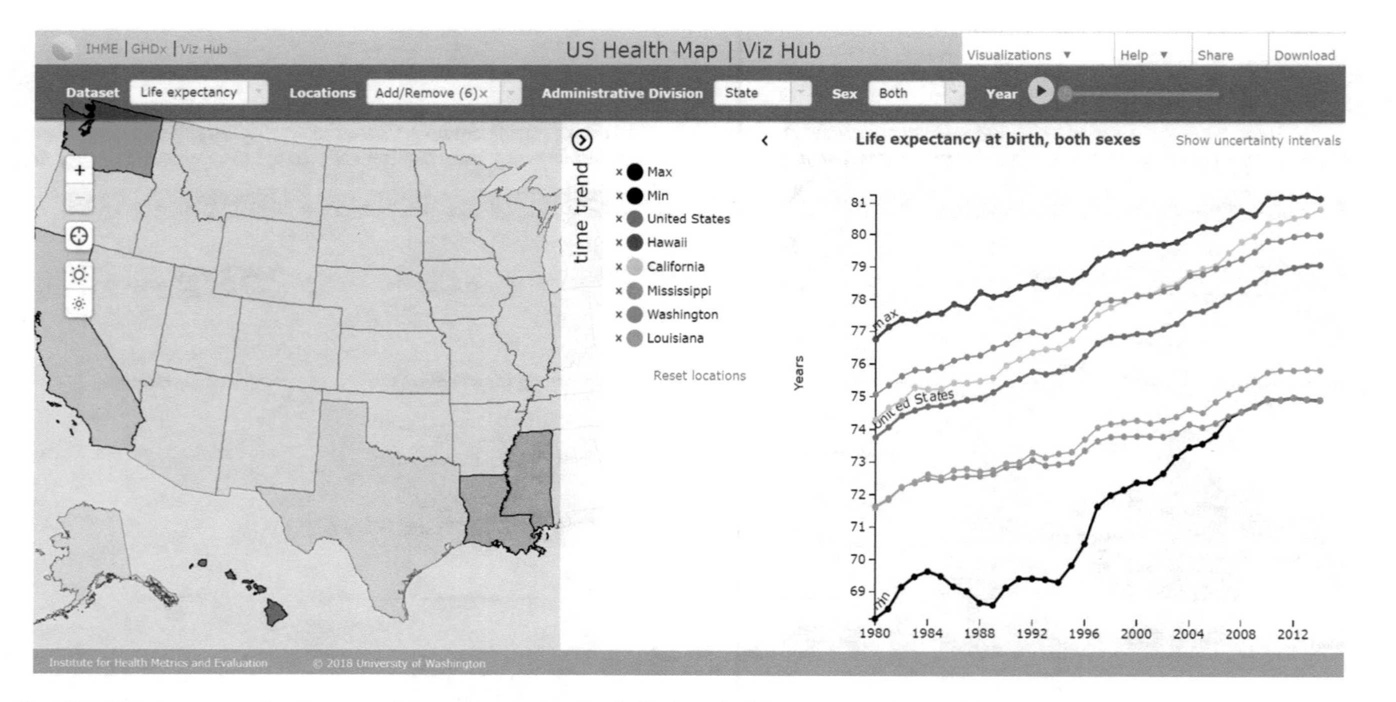

Fig. 5.14 This is an example of a geospatial map and time series. It displays the life expectancy in the United States between 1980 and 2012. It also includes the data for select States, such as Hawaii, which, according to the chart, had the highest life expectancy for more than three decades. The graph was generated on the Health Metrics and Evaluation (IHME) website [9]

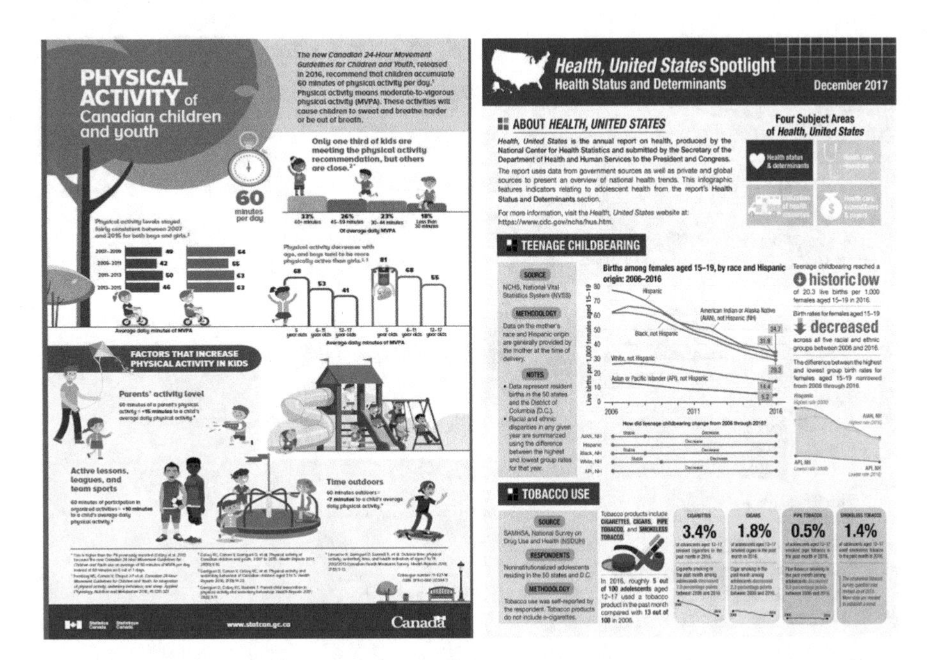

Fig. 5.15 Examples of Infographics (Source: Statistics Canada and National Center for Health Statistics [14, 18])

5.3 Infographics

The term infographics is an abbreviation of "information graphics." They are a combination of data visualizations, text and images, presented in a logical manner similar to storytelling, and are used to convey information and messages in an attractive and easy to understand format [1, 10]. Infographics use many different visual cues to convey information. With the overwhelming amount of data and content generated and shared online, infographics have become very important due to their ability to present information to an audience in a way that can capture and keep the audience's attention, engage it, and aid in the comprehension and retention of the material [11]. Infographics are used in multiple disciplines, such as public policy, journalism, business, and politics. In the healthcare field, infographics are used for health communication and engagement, particularly to support comprehension among individuals with low health literacy [12]. They are helpful tools for communicating key messages clearly, challenging people's thinking, and changing behaviors and attitudes [10]. Figure 5.15 shows two examples of health infographics. The first one was created by the Canadian government (https://www150.statcan.gc.ca/n1/pub/11-627-m/11-627-m2017034-eng.htm) to inform the public of its new 24-h movement guidelines for children and youth [13]. It includes general information, basic statistics about the current physical activity levels in the country, and the factors that can increase them, all in a simple, clear, easy to understand, and visually stimulating fashion for both parents and young people. The second example is an

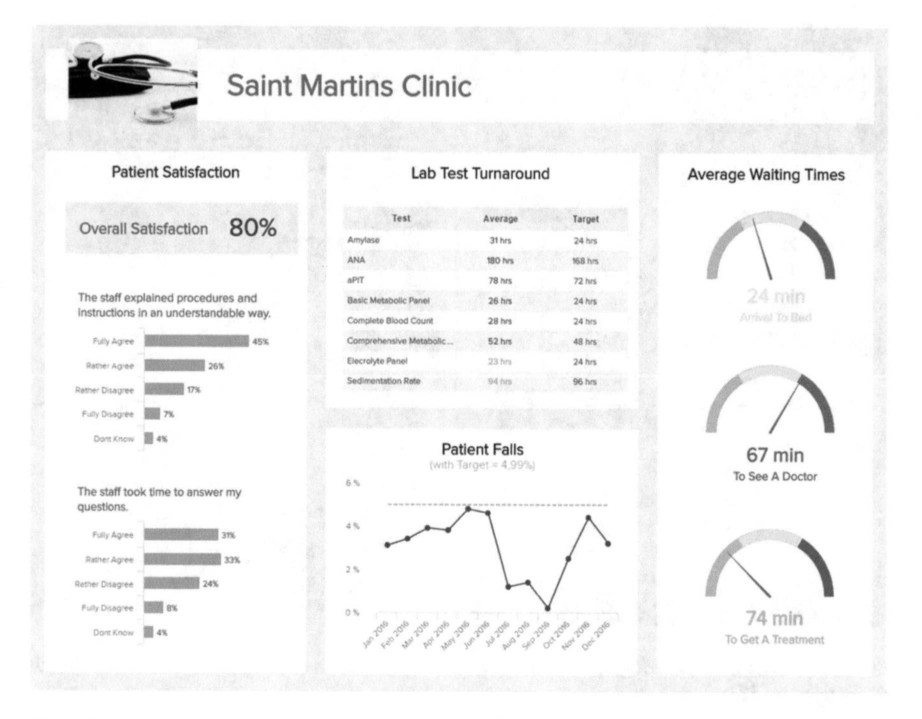

Fig. 5.16 Example of a patient satisfaction dashboard (Source: Datapine.com [19])

infographic of the annual report on health in the United States produced by the National Center for Health Statistics (https://www.cdc.gov/nchs/hus/spotlight/2017-december.htm). It presents an overview of national health trends in the USA and covers topics such as teenage childbearing, tobacco use, suicide deaths, and obesity [14]. Unlike the first example, this infographic is designed and formatted for professionals and politicians.

In general, health infographics are designed without complex medical terminology, allowing the public to understand the message without explanation from health professionals [10]. A study on the design of infographics for engaging community members with varying levels of health literacy found that successful designs are rich in information but without distracting details. They support comparison, between treatments, for example, with a clear recommendation. They provide valuable contextual information and use familiar color and symbolic analogies such as the battery charging level to represent a patient's sleep and energy levels [12].

5.4 Dashboards

The dashboard, which we introduced in Chap. 1, is "a visual display of the most important information needed to achieve one or more objectives, consolidated and arranged on a single screen so the information can be monitored at a glance" [15].

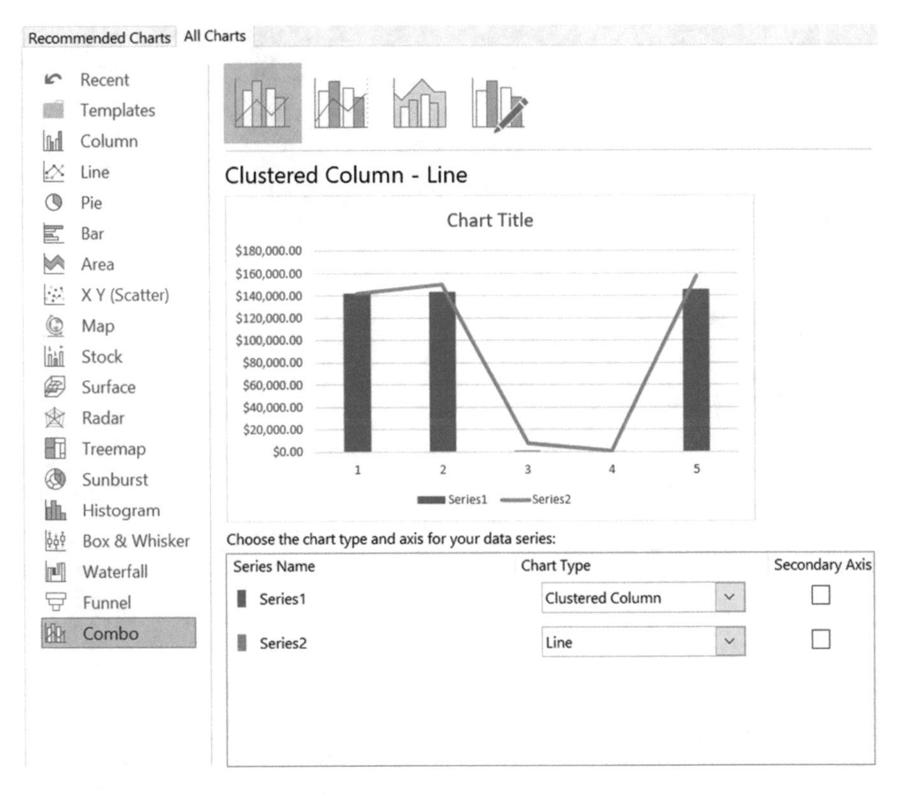

Fig. 5.17 Visualization with Microsoft Excel

Dashboards can be used, for example, to monitor key hospital performance indicators, such as patient satisfaction, average wait times, readmission rates, and average hospital stay (Fig. 5.16). More examples of healthcare dashboards can be found in Chap. 1 of this book and at websites of business intelligence companies such as Datapine (https://www.datapine.com/dashboard-examples-and-templates/healthcare) and Sisense (https://www.sisense.com/dashboard-examples/healthcare/).

Dashboards can be broken down into three roles: strategic, analytical, and operational. At the executive level of an organization, such as a hospital, dashboards support long-term strategic decisions and focus on high-level measures of performance, including future forecasts. They tend to be simple and not interactive and do not require real-time data updates. Dashboards that support data analysis demand rich comparisons, more extensive history, and interaction with data such as drilling down for more details. They can help detect patterns in the data to identify the causes of problems, for example. Similar to the strategic dashboards, they work with static, not real-time, data. Finally, operational dashboards are dynamic and immediate in their nature. They present real-time data in a simple way but also have the means to attract attention in cases when an operation falls outside the range of the acceptable threshold of performance [15].

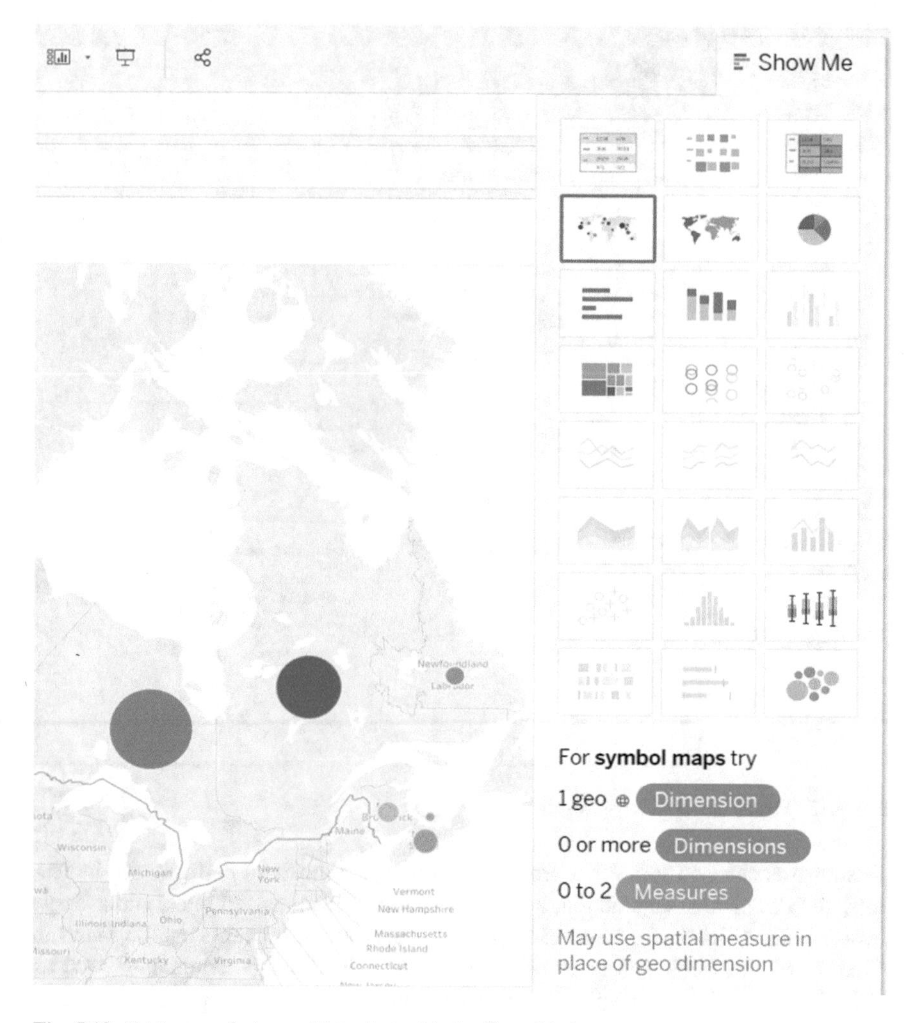

Fig. 5.18 Tableau analytics tool interface with the Show Me button

5.5 Data Visualization Software

The creation of appealing and beautiful visualizations can be achieved with a very large number of software tools starting with the common Microsoft Excel. Today, Excel can create many different basic charts and graphs, such as columns and lines and complex charts and graphs, such as treemaps and waterfalls. It can also create combinations of charts such as clustered columns and lines (Fig. 5.17). For the novice user, Excel recommends the most appropriate charts to use based on the data selected in the spreadsheet.

Business analytics tools such as Tableau Desktop and Lumira by SAP provide a very large number of visualizations that are easy to use and do not require technical knowledge. Most of the work done is via simple point, click, and drag with the mouse. These advanced analytics tools can also interpret the data to identify dimensions and measures, which are comparable to the categorical and quantitative data discussed earlier in the chapter. The charts to use are recommended based on the data available. Tableau Desktop, for example, has a Show Me button (Fig. 5.18) that highlights the available charts that can be used based on the data and makes suggestions for using them. Such tools also allow you to easily create presentations, infographics by combining different charts, and dashboards connected to dynamic data sources.

While we have discussed Tableau, SAP Lumira, and Microsoft Excel in this chapter and used them to generate the visualizations above, it is important to note that there are many large software companies, such as SAS and IBM, and relatively smaller niche players, such as QlikView, that have software with remarkable visualization capabilities. An article in Forbes magazine lists Tableau, QlikView, FusionCharts, Highcharts, Datawrapper, Plotly, and Sisense as the best data visualization tools available in 2017 [16]. Another article by PC Magazine lists Microsoft Power BI, Tableau Desktop, and IBM Watson Analytics as the three best data visualization tools in 2018 [17].

5.6 Conclusion

Data visualization is a critical capability for understanding and interpreting complex data and relationships. Graphs and charts can tell a story, highlight trends, identify outliers and deviations, make comparisons and more, in a simple and effective way. There are many types of graphs and charts available and selecting the one that best matches the data, and the questions you are trying to answer is crucial. Bad visualizations are difficult to understand and can distort what the data are trying to tell us. Today, it is easy to create very rich visualizations using modern analytics tools with simple point and clicks; however, it remains critical to have a good understanding of the data to select the best visualization and be able to interpret it.

References

1. N. Kalé and N. Jones, *Practical Analytics*. Epistemy Press, 2015.
2. S. Few, *Show me the numbers: Designing tables and graphs to enlighten*. Analytics Press, 2012.
3. M. Friendly, "Gallery of Data Visualization," Accessed on: July 18, 2018Available: www.datavis.ca/gallery/
4. CHHS. (2018, 6 August 2018). *Death by ZIP Code by Gender, 2012 - Current*. Available: https://data.chhs.ca.gov/dataset/death-by-zip-code-by-gender-2012-2013
5. CHHS. (2018, 8 August 2018). *West Nile Virus Cases, 2006-present*. Available: https://data.chhs.ca.gov/dataset/west-nile-virus-cases-2006-present

6. S. Canada. (2018, 2018-06-28). *Statistics Canada. Table 13-10-0394-01 Leading causes of death, total population, by age group*. Available: https://www150.statcan.gc.ca/t1/tbl1/en/tv.action?pid=1310039401
7. S. Canada. (2018, 18-08-2018). *Statistics Canada. Table 13-10-0111-02 Cancer, new cases, by selected primary site of cancer and sex*. Available: https://www150.statcan.gc.ca/t1/tbl1/en/tv.action?pid=1310011102
8. S. Canada, "Canadian Community Health Survey, 2012: Annual Component," ed, 2013.
9. IHME. (2018). *Data Visualizations*. Available: http://www.healthdata.org/results/data-visualizations
10. H. Scott, S. Fawkner, C. Oliver, and A. Murray, "Why healthcare professionals should know a little about infographics," ed: BMJ Publishing Group Ltd and British Association of Sport and Exercise Medicine, 2016.
11. J. Lankow, J. Ritchie, and R. Crooks, *Infographics: The power of visual storytelling*. John Wiley & Sons, 2012.
12. A. Arcia *et al.*, "Sometimes more is more: iterative participatory design of infographics for engagement of community members with varying levels of health literacy," *Journal of the American Medical Informatics Association,* vol. 23, no. 1, pp. 174–183, 2015.
13. S. Canada, "Physical activity of Canadian children and youth," ed, 2017.
14. N. C. f. H. Statistics, "Health, United States Spotlight," ed, 2017.
15. S. Few, *Information dashboard design*. O'Reilly, 2006.
16. B. Marr. (2017, 20 July 2017) The 7 Best Data Visualization Tools Available Today. *Forbes*. Available: https://www.forbes.com/sites/bernardmarr/2017/07/20/the-7-best-data-visualization-tools-in-2017/#500386996c30
17. P. Baker. (2018). *The Best Data Visualization Tools of 2018*. Available: https://www.pcmag.com/roundup/346417/the-best-data-visualization-tools
18. Statistics Canada. (2017, September 17). *Physical activity of Canadian children and youth*. Available: https://www150.statcan.gc.ca/n1/pub/11-627-m/11-627-m2017034-eng.htm
19. Datapine.com. (2018, September 17). *Healthcare Dashboards Examples*. Available: https://www.datapine.com/dashboard-examples-and-templates/healthcare#patient-satisfaction-dashboard

Chapter 6
Future Directions

Abstract This concluding chapter focuses on three trends that will affect the future direction of healthcare analytics. Artificial intelligence (AI), which was covered earlier, will be revisited along with the Internet of Things (IoT). The chapter then introduces the concept of big data with its characteristics, known as Vs. The chapter then covers key benefits that are expected from big data analytics in the healthcare industry. The chapter touches upon some ethical concerns, future trends, suggestions for experimenting with healthcare analytics demos, a conclusion, and a list of references.

Keywords Artificial intelligence (AI) · Machine learning · Internet of Things (IoT) · Big data · Ethics

Objectives

At the end of this chapter, you will be able to:

1. Understand the main trends in analytics applications in healthcare
2. Identify the different domains where AI is expected to have an impact
3. Understand the IoT
4. Appreciate the impact of the IoT on healthcare
5. Understand current and future directions of big data analytics in healthcare
6. Identify ethical challenges that these trends have

6.1 Introduction

Healthcare analytics have significantly advanced in the last few years and are expected to continue a trajectory of increased adoption and impact. We will describe three developments in this chapter that will contribute to this growth: artificial intelligence (AI) and machine learning, the Internet of Things (IoT), and big data

C. El Morr, H. Ali-Hassan, *Analytics in Healthcare*, SpringerBriefs in Health Care Management and Economics, https://doi.org/10.1007/978-3-030-04506-7_6

analytics. Each development or trend will have a significant impact on the healthcare industry. AI, for example, is expected to support high-quality and integrated clinical decision-making in the domain of diagnosis. IoT, for example, will facilitate the generation of unprecedented amounts of data that will increase our knowledge and ability to make informed decisions. This unprecedented amount of data is part of what is known as big data. Big data, coupled with advanced analytics, will open the door for new applications such as precision individualized medicine. The three trends and their potential impact will be described in more detail below.

6.2 Artificial Intelligence and Machine Learning Trends

AI aims at mimicking human cognitive abilities and can be applied in a variety of fields in the healthcare field. To illustrate the advances in AI, we can look at a well-known domain of applications: gaming. In May 1997, IBM supercomputer Deep Blue used AI techniques to defeat world champion chess player Gary Kasparov (Fig. 6.1).

Fig. 6.1 Deep blue, By James the photographer [CC BY 2.0 (https://creativecommons.org/licenses/by/2.0)], via Wikimedia Commons

On March 9, 2016, Google AI software, Google DeepMind "AlphaGo," defeated the 17-time world champion Lee Sedol in one of the most complex games ever created, the Go board game [1]. However, in 2017, "AlphaGo Zero," an enhanced version of AlphaGo, not only beat AlphaGo but also learned the game by itself (e.g., unsupervised learning) without data from human-played games [2].

AI techniques such as machine learning algorithms can be used for structured and unstructured healthcare data. Artificial neural networks, deep learning, support vector machines, and natural language processing all have applications in the healthcare field, as we observed in Chap. 4. AI can assist physicians, radiologists, and radiotherapists in making better-informed decisions in the workplace, benefiting patients and enhancing health outcomes.

There have been many advances in AI applications in healthcare to the extent that a debate has arisen regarding the future of radiologists and whether they will be replaced with AI software [3]. During the last few years, AI has been used often in the domain of diagnosis, specifically in diagnosis imaging [4], genetic testing, and electrodiagnosis [5].

AI healthcare applications fall into two categories. The first category uses machine learning techniques to analyze structured data (e.g., images, demographics) to infer the probability of disease outcomes or to perform patient clustering based on certain characteristics [6]. The second category uses NLP to process unstructured data to provide new insights that can add to structured data [7]. Most of the current AI literature focuses on neurology [8], oncology, and cardiovascular diseases. For example, IBM Watson proved to be reliable for oncology [9, 10], deep learning algorithms can precisely identify head CT scan anomalies necessitating urgent attention [11], and the performance of the software based on machine learning systems is comparable to experienced radiologists.

Most likely, we will see the development of new statistical approaches tailored to specific problems in healthcare, such as medical imaging [4]. AI is poised to improve operational performance and efficiency, supporting high-quality and integrated clinical decision-making, enabling population health management, and empowering patients and individuals [12].

6.3 Internet of Things (IoT)

The International Telecommunication Union defines the Internet of Things (IoT) as "a global infrastructure for the information society, enabling advanced services by interconnecting (physical and virtual) things based on existing and evolving interoperable information and communication technologies" [13].

The IoT is the result of many innovations that ultimately created the infrastructure (hardware, software and standards) to enable global seamless connectivity. Apple's invention of the first smartphone (i.e., the iPhone) in 2007 can be considered a crucial event in the IoT pathway [14], as it allowed global individual mobile connectivity and enabled mobile global group connectivity. The total number of smartphone subscriptions reached 4.3 billion in 2017 and is expected to reach 7.2 billion in 2023 according to Ericsson mobility [15].

Additionally, technologies such as radio-frequency identification (RFID) (that enables automatic RFID-tagged object tracking), cloud computing, microbots (micro-robots), and artificial intelligence and machine learning have allowed the collection of unprecedented amounts of data, aggregating, analyze and communicating the resulting information and knowledge on a large scale. The IoT emerged from the integration of these technologies, creating immense opportunities in healthcare and other fields but also immense challenges related to privacy and security and also to health and well-being: a person hacking a pacemaker is a serious health hazard and challenge in an interconnected world of "things." In 2017, the US Food and Drug Administration (FDA) recalled approximately 500,000 pacemakers due to vulnerability to hacking [16].

6.4 Big Data Analytics

Today, we live in the era of big data. There is no unique or universal definition of big data, but there is a general agreement that there has been an explosion of data generation, storage, and usage [17]. Big data is a popular term used to describe the exponential growth, availability, and use of information, both structured and unstructured [18]. These data come from daily business transactions at banks and retailers, for example, from sensors such as security cameras and monitoring systems, from GPS systems on every mobile phone, from content posted on social media such as YouTube videos, and from many more ubiquitous sources. Big data in the healthcare field comes from medical devices such as MRI scanners and X-ray machines, sensors such as heart monitors, patient electronic medical and health records, insurance providers' records, doctors' notes, genomic research studies, wearable devices, and many more [19]. In August 2018, it was announced that Fitbit, a manufacturer of wearable activity trackers, had collected 150 billion hours' worth of heart rate data from tens of millions of people from all over the world. These data also include sex, age, locations, height, weight, activity levels, and sleep patterns. Moreover, Fitbit has 6 billion nights' worth of sleep data [20].

There are multiple factors behind the emergence and growth of big data, and they include technological advances in the field of information and communication technology (ICT), where computing power and data storage capacity are continuously increasing while their cost is decreasing. The increased connectivity to the Internet is another major factor. Today, most people have a mobile device, and many modern pieces of equipment are connected to the Internet.

Big data is generally characterized by the 4 Vs: volume, variety, velocity (introduced originally by the Gartner Group in 2001), and veracity (added later by IBM) [21]. Multiple additional Vs were introduced later, including validity, viability, variability, vulnerability, visualization, volatility, and value [17, 21, 22]. We will describe the key terms below. Volume is the most defining characteristic of big data. The

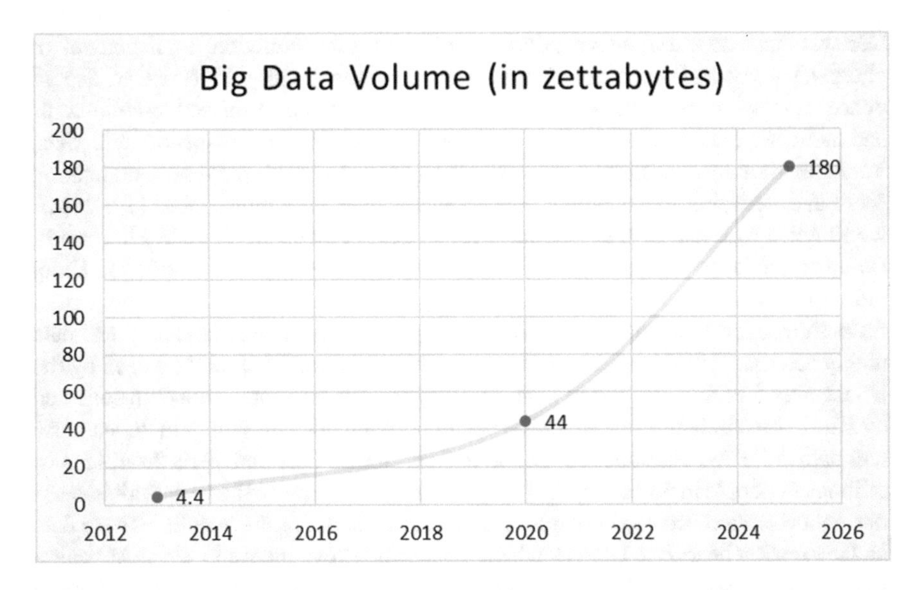

Fig. 6.2 Exponential increase in the volume of big data (actual and projected) based on an IDC study

volume of data generated is increasing exponentially, and new units of measure have been created, such as zettabytes (10^{21}), to accommodate this increasing volume of data. According to IDC, a market-research firm, the data created and copied in 2013 was 4.4 zettabytes, and this number is projected to exponentially increase to 44 zettabytes in 2020 and 180 zettabytes in 2025 (Fig. 6.2) [21, 23]. Examples of large volumes of data are the 20 terabytes (10^{12}) of data produced by Boeing jets every hour and the one terabyte of data that is uploaded on YouTube every 4 min [24].

Variety refers to the different forms of big data, such as text-in patient medical records, images from X-ray machines, videos from MRI scanners, location data from GPS systems, and other formats, for example, from wearable wireless health monitors. Velocity refers to the very high speed at which big data are continuously being generated, for example, from medical devices and monitors in hospitals' intensive care units. It is critical for such data to be generated and analyzed in real time. Finally, veracity represents the high level of uncertainty and low levels of reliability and truthfulness of big data [17, 21, 25]. Data can be biased, incomplete, or filled with noise; indeed, healthcare data scientists and analysts spend more than 60% of their time cleaning the data [21]. These characteristics of big data represent challenges for any company or industry. Some of the challenges are technical, such as being able to analyze the large volume of data, generating very rapidly, and in many different formats. Other challenges may be administrative, such as the reliability of the data. Nevertheless, big data analytics provide many opportunities for the healthcare industry.

Big data analytics are expected to significantly improve healthcare benefits and reduce costs. Among the major potential advantages of big data analytics are personalized healthcare to identify best-fit and cost-effective treatment, predictive disease management and early intervention, readmission rate reduction, management of pharmacy costs and outcome, identification of patients with a high risk of dependence on drugs or developing chronic diseases, prediction of missed appointments and noncompliance with medication based on health and socioeconomic data, improved monitoring, improved coordination between health providers, combating fraud and verifying the accuracy of insurance claims, and many more [19, 26]. A well-known example of big data analytics in healthcare is the Google Flu Trends. Based on the large volume of data from individuals searching for information about influenza when they are sick, Google is able to detect trends and estimate the current flu outbreak levels in the USA and other regions. Not only is Google's estimate highly accurate, but it is also 2 weeks faster than the traditional method used by the US Centers for Disease Control [27]. Another big data example is the data collected by Fitbit, a manufacturer of wearable health tracking devices, which includes pulse rate data collected continuously day and night, for months and years from tens of millions of people in 55 countries. These data are in contrast to the very limited heart rate data collected occasionally during visits to the doctor or the hospital. These data include resting heart rate (RHR), which is an informative metric in terms of health, fitness, and lifestyle that has been linked to early death and diabetes. The Fitbit big data have shown an association between high RHR and low body weight. It also revealed, contrary to popular belief, that the positive effect of exercise on RHR tapers off after a couple hundred minutes of exercise per week and does not continue to increase, that exercise can lower RHR at any age, including the elderly and that 7.25 h of sleep, and not 8 h as previously thought, are optimal for heart health. Finally, the data show a variation in the outcomes based on the country where the individual is from [20].

6.5 Ethical Concerns

The ability of information technology to collect, analyze, transfer, and store massive amounts of health-related data has given rise to several concerns. The first major concern is the risk to data security from hackers and data breaches. In the past few years, healthcare data breaches have grown in size and frequency, with the largest breach affecting almost 80 million people. The exposed data were highly sensitive and included patients' identifying information, health insurance information, and medical histories [28]. In addition to hacking databases and stealing data, criminals may be able to remotely access certain medical devices, such as implantable cardiac pacemakers, if they have IoT capabilities but lacks cybersecurity and cause serious physical harm to patients [16]. Another major concern is related to privacy and the risk that sensitive personal information may be improperly accessed or shared. The personal records of hundreds of high school students in Australia were mistakenly published on their school's intranet, including medical and mental health conditions, learning disabilities, behavioral difficulties, and medications used [29]. Other ethical concerns and questions will arise with advances in healthcare analytics. For example, can insurance companies deny coverage for individuals whose data predict that they will develop certain illnesses or behavioral problems in the future? Who will be held responsible if artificial intelligence makes erroneous decisions? These and other questions do not have a clear answer. While the benefits of healthcare data analytics are evident and their adoption will continue to increase, it is important to find the proper safeguards and policies to protect the data and ensure its ethical use.

6.6 Future Directions

In this section, we list some healthcare analytics trends that experts in the field predict will occur in the near future. Artificial intelligence, machine learning, and predictive analytics will provide more insights into the population health data, such as the financial implications of treatment protocols. Better accountability is expected because data from different health providers are aggregated on scorecards, powered by visual dashboards, for example, and will measure physician performance and utilization and patient satisfaction and will enable improvements. Easy-to-use self-serve analytics tools will empower healthcare professionals such as physicians, clinicians, and nurses to find answers to questions and be able to make data-driven decisions [30]. Big data coming from mobile biometric sensors, smartphone apps, and genomics will further enable precision individualized medicine as opposed to the traditional one-size-fits-all types of treatment. Better patient profiles and predictive models will enhance the anticipation, diagnosis, and treatment of diseases. Real-time analytics will help identify early signs of infections such as sepsis. Finally, patient data predictive analytics will help cut healthcare costs by reducing the rate of hospital readmission, forecast operating room demand, optimize staffing and streamline patient care [31].

6.7 Healthcare Analytics Demos

In this final chapter, we suggest you try first a demo of healthcare analytics applications from Qlik. Go to https://www.qlik.com/us/solutions/industries/healthcare and click on the Try Demo button in the Improve Quality of Care section. You will then get the chance to experiment with the dashboard of seven different applications such as patient analysis and bed day analysis. Select different options from the drop-down menus and see how the output changes based on your selection. You can explore additional healthcare analytics dashboards provided by Sisense at https://www.sisense.com/glossary/healthcare-analytics-basics/

Another analytics environment worth exploring is IBM Watson at https://www.ibm.com/watson-analytics. Signup for a free account and experiment with its discovery and predictive analytics capabilities. The system is very intuitive and includes videos that show you how to load your data, discover insights by asking questions in plain English, generate instant visualizations, perform predictive analytics, and create a basic dashboard.

6.8 Conclusion

New growth in technologies is changing the way we process data and make decisions. Like other technologies, AI, the IoT, and big data are here to stay; they provide immense benefits in healthcare as we discovered in Chap. 4; however, they raise immense ethical concerns in healthcare and beyond. They also raise societal issues, for example, a mathematician, Cathy O'Neil, felt compelled to write a book to address the impact of big data on inequality and democracy [32]; the book's name ("Weapons of Math Destruction") is reflective of a general human acknowledgment and concern about these changes and a call to reflect on them.

Analytics in healthcare have immense benefits and will change the way we deliver care for people. Professionals in some fields are feeling the change first; radiologists, for instance, are raising questions about the future of radiology and radiography [33]; an AI software that is able to read an image better than a trained human is not a farfetched idea anymore, and a robot that can take an X-ray in a more precise way than humans can might be feasible. If radiographers and radiologists do not disappear, their work will undergo an immense change; other fields will probably face similar challenges, and society as a whole needs to make decisions, and citizens need to weigh in about the directions these technologies should be taking and to what extent. Ultimately, if something can be created, it does not necessarily mean that it ought to be created; our future should be decided by our personal and collective efforts and reflection that translates into policies.

References

1. C. Sang-Hun and J. Markoff. (2016, August 31). *Master of Go Board Game Is Walloped by Google Computer Program.* Available: https://www.nytimes.com/2016/03/10/world/asia/google-alphago-lee-se-dol.html
2. I. Sample. (2017, August 31). *"It's able to create knowledge itself": Google unveils AI that learns on its own.* Available: https://www.theguardian.com/science/2017/oct/18/its-able-to-create-knowledge-itself-google-unveils-ai-learns-all-on-its-own
3. The Economist. (2018, August 31). *Images aren't everything: AI, radiology and the future of work.* Available: https://www.economist.com/leaders/2018/06/07/ai-radiology-and-the-future-of-work
4. S. Wang and R. M. Summers, "Machine learning and radiology," (in eng), *Med Image Anal,* vol. 16, no. 5, pp. 933–51, Jul 2012.
5. F. Jiang *et al.*, "Artificial intelligence in healthcare: past, present and future," (in eng), *Stroke Vasc Neurol,* vol. 2, no. 4, pp. 230–243, Dec 2017.
6. A. M. Darcy, A. K. Louie, and L. Roberts, "Machine learning and the profession of medicine," *JAMA,* vol. 315, no. 6, pp. 551–552, 2016.

7. H. J. Murff, F. FitzHenry, M. E. Matheny, and et al., "Automated identification of postoperative complications within an electronic medical record using natural language processing," *JAMA,* vol. 306, no. 8, pp. 848–855, 2011.
8. C. E. Bouton *et al.*, "Restoring cortical control of functional movement in a human with quadriplegia," *Nature,* vol. 533, p. 247, 04/13/online 2016.
9. Y. Y. Kim, S. J. Oh, Y. S. Chun, W. K. Lee, and H. K. Park, "Gene expression assay and Watson for Oncology for optimization of treatment in ER-positive, HER2-negative breast cancer," (in eng), *PLoS One,* vol. 13, no. 7, p. e0200100, 2018.
10. S. P. Somashekhar *et al.*, "Watson for Oncology and breast cancer treatment recommendations: agreement with an expert multidisciplinary tumor board," (in eng), *Ann Oncol,* vol. 29, no. 2, pp. 418–423, Feb 1 2018.
11. S. Chilamkurthy *et al.*, "Development and Validation of Deep Learning Algorithms for Detection of Critical Findings in Head CT Scans," vol. abs/1803.05854, 2018.
12. Philips.com. (2018). *Using AI to meet operational clinical goals*. Available: https://www.philips.co.uk/c-dam/b2bhc/master/seamless-care/Q1-HIM/AI_Updated_02032018.pdf
13. ITU Telecommunication Standardization Sector. (2012, August 30, 2018). Available: https://www.itu.int/ITU-T/recommendations/rec.aspx?rec=y.2060
14. S. Greengard, *The Internet of Things* (MIT Press Essential Knowledge series). MIT Press, 2015.
15. Ericsson Mobility. (2018, August 31). *Ericsson Mobility Report*. Available: https://www.ericsson.com/assets/local/mobility-report/documents/2018/ericsson-mobility-report-june-2018.pdf
16. A. Hern. (2017, August 31). *Hacking risk leads to recall of 500,000 pacemakers due to patient death fears*. Available: https://www.theguardian.com/technology/2017/aug/31/hacking-risk-recall-pacemakers-patient-death-fears-fda-firmware-update
17. N. Kalé and N. Jones, *Practical Analytics*. Epistemy Press, 2015.
18. SAS. (2018). *Big Data - What it is and why it matters*. Available: https://www.sas.com/en_ca/insights/big-data/what-is-big-data.html
19. D. Faggella, "Where Healthcare's Big Data Actually Comes From," 11 January 2018. Available: https://www.techemergence.com/where-healthcares-big-data-actually-comes-from/
20. D. Pogue, "Exclusive: Fitbit's 150 billion hours of heart data reveal secrets about health," August 27, 2018. Available: https://finance.yahoo.com/news/exclusive-fitbits-150-billion-hours-heart-data-reveals-secrets-human-health-133124215.html?linkId=56096180
21. J. Bresnick, "Understanding the Many V's of Healthcare Big Data Analytics," 5 June 2017. Available: https://healthitanalytics.com/news/understanding-the-many-vs-of-healthcare-big-data-analytics
22. R. Sharda, D. Delen, and E. Turban, *Business Intelligence: A managerial Perspective on Analytics*. Prentice Hall Press, 2015.
23. T. Economist. (2017, 6 May) Data is giving rise to a new economy. *The Economist*.
24. R. Sharda, D. Delen, E. Turban, J. Aronson, and T. P. Liang, *Businesss Intelligence and Analytics: Systems for Decision Support*. 2014.
25. R. Sharda, D. Delen, and E. Turban, *Business Intelligence: A Managerial Perspective on Analytics: A Managerial Perspective on Analytics*. Pearson, 2015, pp. 416–416.
26. A.-L. Beall, "Big data in health care," Accessed on: 3 September 2018 Available: https://www.sas.com/en_ca/insights/articles/big-data/big-data-in-healthcare.html
27. M. Karch, "Track Influenza With Google Flu Trends," Accessed on: 3 September 2018
28. N. Lord, "Top 10 Biggest Healthcare Data Breaches of All Time," *Data Insider*, 25 June 2018 Available: https://digitalguardian.com/blog/top-10-biggest-healthcare-data-breaches-all-time
29. Australian Associated Press, "Melbourne student health records posted online in 'appalling' privacy breach," in *The Guardian*, ed, 2018.
30. HealthData Management, "5 trends in healthcare analytics for 2018," 5 February 2018. Available: https://www.healthdatamanagement.com/list/tableau-software-assesses-coming-changes-in-healthcare-analytics

31. J. V. Wagenen, "3 Big Data Trends in Healthcare Using Predictive Analytics," 28 November 2017. Available: https://healthtechmagazine.net/article/2017/11/predicting-analytics-3-big-data-trends-healthcare
32. C. O'Neil, *Weapons of Math Destruction: How Big Data Increases Inequality and Threatens Democracy*. Crown/Archetype, 2016.
33. O. A. Paiva and L. M. Prevedello, "The potential impact of artificial intelligence in radiology," *Radiologia Brasileira,* vol. 50, no. 5, pp. V-VI, Sep-Oct 2017.

Index

C. El Morr, H. Ali-Hassan, *Analytics in Healthcare*, SpringerBriefs in Health Care Management and Economics, https://doi.org/10.1007/978-3-030-04506-7

Printed in the United States
By Bookmasters